SpringerBriefs in Physics

SpringerBriefs in Physics are a series of slim high-quality publications encompassing the entire spectrum of physics. Manuscripts for SpringerBriefs in Physics will be evaluated by Springer and by members of the Editorial Board. Proposals and other communication should be sent to your Publishing Editors at Springer.

Featuring compact volumes of 50 to 125 pages (approximately 20,000–45,000 words), Briefs are shorter than a conventional book but longer than a journal article. Thus, Briefs serve as timely, concise tools for students, researchers, and professionals.

Typical texts for publication might include:

- A snapshot review of the current state of a hot or emerging field
- A concise introduction to core concepts that students must understand in order to make independent contributions
- An extended research report giving more details and discussion than is possible in a conventional journal article
- A manual describing underlying principles and best practices for an experimental technique
- An essay exploring new ideas within physics, related philosophical issues, or broader topics such as science and society

Briefs allow authors to present their ideas and readers to absorb them with minimal time investment. Briefs will be published as part of Springer's eBook collection, with millions of users worldwide. In addition, they will be available, just like other books, for individual print and electronic purchase. Briefs are characterized by fast, global electronic dissemination, straightforward publishing agreements, easy-to-use manuscript preparation and formatting guidelines, and expedited production schedules. We aim for publication 8–12 weeks after acceptance.

More information about this series at https://link.springer.com/bookseries/8902

Ioachim Pupeza

Passive Optical Resonators for Next-Generation Attosecond Metrology

 Springer

Ioachim Pupeza
Max-Planck-Institut für Quantenoptik
Munich, Bayern, Germany

ISSN 2191-5423   ISSN 2191-5431   (electronic)
SpringerBriefs in Physics
ISBN 978-3-030-92971-8   ISBN 978-3-030-92972-5   (eBook)
https://doi.org/10.1007/978-3-030-92972-5

This Springer imprint is published by the registered company Springer Nature Switzerland AG
The registered company address is: Gewerbestrasse 11, 6330 Cham, Switzerland

To my family

Preface

The unparalleled control over broadband light fields afforded by laser architectures based on visible/near-infrared (VIS/NIR) phase-stabilized mode-locked oscillators has steadily spawned new means for expanding our understanding of basic processes in nature and the universe. On the one hand, the temporal confinement of light to bursts with durations reaching down to merely a few oscillations of the optical carrier wave, with exquisite repeatability and with field strengths rivaling those of the atomic Coulomb fields, has enabled real-time measurements (and control) of ultrafast processes with ever-improving temporal resolution. Nowadays, *attosecond metrology* grants experimental access to the fastest events outside the atomic core, namely transitions between quantum electronic states that determine the physical and chemical properties of atoms, molecules, and condensed matter and mediate chemical reactions or biological processes.

On the other hand, the temporal periodicity of the pulse train emitted by mode-locked oscillators results in a comb-like spectrum, consisting of equidistant laser lines separated by the pulse repetition frequency. *Frequency combs* thus conveniently link the domains of optical frequencies (few-to-100s of THz) and of radio frequencies (~100 MHz) accessible to contemporary electronics. This enables the most precise measurements of time with optical clocks permitting, e.g., the search for miniscule drifts of natural constants.

The availability of radiation sources providing broadband coverage of the vacuum-/extreme-ultraviolet spectral ranges (henceforth, for simplicity referred to as XUV) with multi-MHz pulse repetition frequencies and coherence properties akin to those of mode-locked oscillators promises a substantial impact on precision frequency metrology, photoelectron spectroscopy (PES) and attosecond science. For instance, a particularly severe shortcoming of laser systems employed in attosecond-resolution PES arises from their relatively low pulse repetition frequencies, usually in the lower kHz range, in the context of space charge effects limiting the admissible number of photoelectrons ejected during each laser pulse from a solid. At typical nanosecond-range travel times of photoelectrons inside the detector, the detection duty cycle in state-of-the-art measurements amounts to a fraction of a percent, resulting in unpractically or even prohibitively long measurement times. In addition, applications like precision spectroscopy of hydrogen-like ions for tests of bound-state quantum electrodynamics, or the realization of robust, multi-PHz optical clocks tuned to nuclear transitions have constituted another important impetus for the development of XUV frequency combs.

This book reviews the research activities conducted by the author and his team over the course of five years toward such a source and its experimental validation. Primarily motivated by the development of a new generation of high-repetition frequency instruments for multi-dimensional attosecond PES, the source, based on cavity-enhanced high-harmonic generation (HHG), operates at a repetition frequency of 18.4 MHz, emitting 5×10^5 photons per pulse in the 25–60 eV range. The emission is scalable to higher photon energies (> 100 eV) at the cost of the photon rate. This book summarizes and arranges the developments leading to this source, as well as proof-of-principle ultrafast attosecond PES measurements, published in 18 original publications, and is structured as follows.

As a first prerequisite for efficient intracavity HHG, power scaling of femtosecond enhancement cavities is thoroughly investigated. Advanced resonator designs are derived and verified, featuring large illuminated spots on all mirrors, mitigating both intensity and thermally induced enhancement limitations. With suitable mirror-substrate and dielectric-coating materials, at a repetition frequency of 250 MHz, record average powers for ultrafast laser technology of 670, 400, and 20 kW are demonstrated for 10-ps, 250-fs, and 30-fs pulses, respectively, circulating in a passive optical resonator.

Secondly, the dynamics of a high-finesse, passive resonator in the presence of a highly nonlinear optical process, such as HHG, are quantitatively investigated in theory and experiment. These investigations were instrumental in allowing for a holistic optimization of the XUV source reported here, which for the first time reached intracavity HHG conversion efficiencies similar to those achieved in single-pass setups with similar gas targets ($\sim 10^{-7}$ in an argon gas target).

Thirdly, we extensively studied geometric XUV output coupling from the enhancement cavity, where the high-photon energy beam exits the cavity through an opening in the mirror following the HHG focus. Among various output coupling methods, geometric output coupling exhibits the advantages of robustness, low distortion to the participating pulses, and photon-energy scalability. Several implementations are discussed in this book, using both the fundamental transverse Gaussian resonator mode and tailored, higher-order modes. The latter offers the prospect of broadband output coupling efficiencies approaching unity, as well as of novel spatio-temporal gating methods for the generation of isolated XUV attosecond pulses.

Finally, we report proof-of-principle attosecond angle-resolved PES experiments carried out at 18.4 MHz, with attosecond pulse trains emerging via HHG driven by sub-40-fs, near-infrared pulses circulating in a femtosecond enhancement cavity. In our experiment, 1×10^{10} photoelectrons were released per second from a 10-μm-diameter spot on tungsten, at space charge distortions of only a few tens of meV. Broadband, time-of-flight photoelectron detection with nearly 100% temporal duty cycle evidences a count rate improvement between two and three orders of magnitude over state-of-the-art attosecond PES experiments under identical space charge conditions, reducing the total measurement time from days to minutes.

The book concludes with an outlook on possible further developments of cavity-enhanced HHG. Firstly, we report the demonstration of temporal dissipative solitons in free-space passive resonators. With the characteristic properties of temporal self-compression and self-stabilization, this novel implementation of femtosecond enhancement cavities might constitute both an efficient way of mitigating nonlinear dynamics-induced intensity limitations in cavity-enhanced HHG and a decisive step toward decreasing the system complexity by combining nonlinear pulse compression and HHG in a single, compact cavity. Cavity solitons might also impact further applications in the future, such as other nonlinear conversion processes, (e.g., the generation of broadband, coherent infrared radiation via optical rectification) or linear and nonlinear cavity-enhanced spectroscopies. Furthermore, we discuss the generation of isolated XUV attosecond pulses via an intracavity wavefront rotation, similar to an "attosecond lighthouse," having demonstrated all necessary prerequisites.

Munich, Germany   Ioachim Pupeza

Contents

1 Introduction 1
  1.1 Tracking Electron Dynamics on Their Native Time
     Scale—Opportunities and Challenges 1
  1.2 Photoemission-Spectroscopy-Based Attosecond Metrology 5
    1.2.1 Photoelectron Spectroscopy—A Brief Historical
       Overview 5
    1.2.2 Attosecond Metrology Based on Photoelectron
       Spectroscopy 6
    1.2.3 Femtosecond Enhancement Cavities 8
    1.2.4 Experimental Implementations and Typical Parameters
       for Cavity-Enhanced HHG 10
  1.3 Research Objectives and Structure of the Book 11
    1.3.1 The Project MEGAS (MHz Attosecond Pulses
       for Photoelectron Spectroscopy and Microscopy) 11
    1.3.2 Structure of the Book 13
  References 15

2 Cavity-Enhanced High-Order Harmonic Generation
  for Attosecond Metrology 19
  2.1 Power Scaling of Femtosecond Enhancement Cavities 19
    2.1.1 Large-Mode Enhancement Cavities [2] 19
    2.1.2 Megawatt-Scale Average-Power Ultrashort Pulses
       in an Enhancement Cavity [4] 21
    2.1.3 Balancing of Thermal Lenses in Enhancement Cavities
       with Transmissive Elements [7] 22
  2.2 Femtosecond Enhancement Cavities in the Nonlinear Regime [8] ... 23
  2.3 Geometric Output Coupling of Intracavity Generated
    High-Order Harmonics 25
    2.3.1 Compact High-Repetition-Rate Source of Coherent 100
       eV Radiation [25] 26

2.3.2   High-Harmonic Generation at 250 MHz with Photon
        Energies Exceeding 100 eV [6]   26
2.3.3   Cavity-Enhanced High-Harmonic Generation
        with Spatially Tailored Driving Fields [29]   29
2.3.4   Cavity-Enhanced Noncollinear High-Harmonic
        Generation [39]   30
2.4 The MEGAS Beamline   32
2.4.1   Phase-Stable, Multi-μJ Femtosecond Pulses
        from a Repetition-Rate Tunable Ti:Sa-Oscillator-Seeded
        Yb-Fiber Amplifier [41]   37
2.4.2   Cumulative Plasma Effects in Cavity-Enhanced
        High-Order Harmonic Generation in Gases [46]   38
2.4.3   Efficiency of Cavity-Enhanced High-Harmonic
        Generation with Geometric Output Coupling [40]   41
2.5 High-Flux Ultrafast Extreme-Ultraviolet Photoemission
     Spectroscopy at 18.4 MHz Pulse Repetition Rate [45]   41
2.5.1   HHG Source   41
2.5.2   Laser-Assisted Photoemission Electron Spectroscopy
        at 18.4 MHz—Photoelectron Statistics   43
2.5.3   Attosecond Angle-Resolved Photoemission Electron
        Spectroscopy (Attosecond-ARPES) at 18.4 MHz   45
References   49
3 Next-Generation Enhancement Cavities for Attosecond
   Metrology—An Outlook   53
3.1 Passive Enhancement of Few-Cycle, Waveform-Stable Pulses   54
3.1.1   Enhancement Cavities for Zero-Offset-Frequency Pulse
        Trains [4]   54
3.1.2   Enhancement Cavities for Few-Cycle Pulses [5]   54
3.2 Toward Intracavity Gating for the Generation of Isolated
     Attosecond Pulses   55
3.2.1   Generation of Isolated Attosecond Pulses
        with Enhancement Cavities—A Theoretical Study [8]   56
3.2.2   Tailoring the Transverse Mode of a High-Finesse Optical
        Resonator with Stepped Mirrors [10]   56
3.2.3   Cavity-Enhanced Noncollinear High-Harmonic
        Generation [11]   58
3.2.4   Interferometric Delay Tracking for Low-Noise Mach–
        Zehnder-Type Scanning Measurements [12]   59
3.3 Solitons in Free-Space Femtosecond Enhancement Cavities [13] ...   60
References   61

Chapter 1
Introduction

1.1 Tracking Electron Dynamics on Their Native Time Scale—Opportunities and Challenges

In our contemporary understanding of the structure and dynamics of matter, electrons play a pivotal role. Quantum electronic states and transitions between them determine the physical and chemical properties of atoms, molecules and condensed matter, they mediate chemical reactions and biological processes. Consequently, advancing our tools for the study of electronic structure and motion on the atomic scale opens up opportunities for deepening our understanding of basic processes in nature and developing novel technologies.

A device capable of accessing these dynamics on their native, attosecond-to-picosecond timescales[1] fundamentally requires sampling within time windows that are shorter than the fastest feature to be observed. The key to ever-increasing temporal resolution has been the outstanding confinement of optical energy to flashes with durations well below 1 ps—in particular, well below the temporal resolution of contemporary electronics—together with highly reproducible shot-to-shot intensity envelopes, such as uniquely afforded by modelocked lasers [1]. Femtosecond visible/infrared pulses have, for instance, enabled studies of the dynamics of chemical bonds at the level of nuclear motion, establishing the field of femtochemistry [2] at the end of the last century.

Around the turn of the century, further developments in femtosecond laser technology [3] brought about the most recent qualitative leap in temporal resolution, opening the door to observations of atomic-scale dynamic processes [4, 5]. The amplification of femtosecond light pulses, their nonlinear spectral broadening and subsequent temporal compression afforded optical transients comprising merely a few oscillations of the carrier electric field under its intensity envelope, with peak

---

[1] In Bohr's model of the hydrogen atom, a revolution of the electron around the proton takes about 150 attoseconds.

© The Author(s), under exclusive license to Springer Nature Switzerland AG 2022
I. Pupeza, *Passive Optical Resonators for Next-Generation Attosecond Metrology*,
SpringerBriefs in Physics, https://doi.org/10.1007/978-3-030-92972-5_1

field strengths exceeding the binding Coulomb fields of atoms. The extremely fast rise of the optical electric field to such high strengths, together with its extremely fast oscillating nature, allowed for the control of the ionization-recollision dynamics underlying high-order harmonic generation (HHG) in gases [6–8].

The radiation generated by femtosecond-laser-driven HHG uniquely combines two features. Firstly, the energies of emitted photons extend from a few eV to the keV range [9], granting access to an ultrabroad range of electronic transitions—including even those from tightly-bound core levels—via single-photon absorption. Secondly, this radiation emerges in ultrashort, typically sub-femtosecond, bursts rigidly locked to the oscillations of the electric field driving HHG. The latter permits attosecond-precision determination of the timing in experiments employing both fields, ultimately enabling insights into the fastest hitherto observable (and controllable) dynamics in nature, namely electron motion within matter.

In the past two decades, several time-resolved spectroscopic techniques with temporal resolution down to the attosecond regime, involving HHG, have emerged. Typically, ultrafast dynamics are initiated in a sample by one of the visible/infrared pulse driving HHG, or the extreme-ultraviolet high-harmonic pulse. The temporal evolution of these dynamics is then probed in various ways [10].

For instance, *transient absorption spectroscopy* follows the absorption of HHG radiation transmitted through the excited sample at well-defined delays with respect to an ultrafast perturbative or strong-field excitation [5]. Importantly, this all-optical technique simultaneously provides high temporal (attosecond) and energy (meV) resolution [5]. This combination does not violate Heisenberg's uncertainty principle because the measurement of the absorption spectrum for a given delay between the pump and the probe pulse is not time-resolved [11].

Time-resolved *photoelectron spectroscopy* (PES[2]) techniques measure electrons released from a sample harnessing the brevity of the involved pump and probe pulses for accurate temporal referencing. Besides providing the perhaps most direct experimental access to electronic wavefunctions, PES affords fundamental advantages over all-optical spectroscopies, its spatial resolution being limited by the de Broglie wavelength of the detected electrons. For example, at photoelectron kinetic energies around 100 eV (nowadays readily achievable with HHG sources around the world) the de Broglie wavelength becomes comparable to the Bohr radius[3] while, in comparison, the wavelength of a photon with a similar energy is roughly two orders of magnitude longer.

High-harmonic generation spectroscopy [12] is another prominent example of an attosecond metrology technique relying on HHG. Here, the HHG process is

---

[2] Often, "PES" denotes "photoelectron spectroscopy." Which definition is more appropriate depends on whether the process of photoemission, or the particle/wave constitutes the focus of the study.

[3] At a photoelectron kinetic energy of 60 eV, achieved with the MHz-repetition-rate HHG beamline presented in this work, the de Broglie wavelength is 1.6×10^{-10} m, which is roughly three times the Bohr radius.

driven in the sample material itself and the observable is the high-harmonic spectrum. Attosecond temporal resolution can be obtained, e.g., by slightly varying the excitation field [13].

The high intensities (on the order of 10^{14} W/cm^2), which are necessary to drive HHG in gases, normally require a substantial reduction of the pulse repetition rate with respect to that of the modelocked oscillator providing the initial pulses. The resulting, relatively low repetition rates (usually in the lower kHz range) bring about particularly severe limitations in state-of-the-art attosecond-PES experiments in the context of space charge [14]. For comparison, 3rd-generation synchrotrons emit pulsed radiation with repetition rates of several hundreds of MHz, set by the temporal spacing of the electron bunches. Coulomb interaction of multiple photoelectrons released from the sample during a single laser pulse affects their velocities and trajectories towards the detector, distorting the observables [15, 16]. In most state-of-the-art setups for multidimensional (i.e., energy, and angularly or spatially resolved) attosecond-PES on solids, space charge effects demand a significant attenuation of the photon flux available from the source for sample illumination [17–19] (Fig. 1.1).

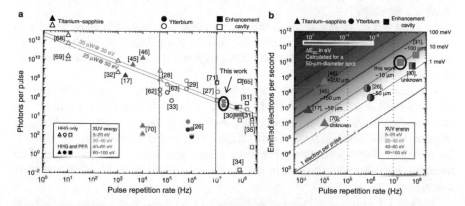

Fig. 1.1 **a** Generated XUV photons per pulse versus pulse repetition rate for different energy ranges (color-coded, see legend) for a representative selection of HHG sources. The shape of the symbols (top) indicates the underlying laser technology. Full/empty symbols: HHG sources which have/have not been used for PES, respectively. Diagonal lines: constant XUV flux at a given energy: 30 eV (orange) and 50 eV (green). Data taken from Refs. [17, 20–39] in Saule et al. [18] (which is referred to as "this work" in the figure). **b** Photoelectron flux and estimated space-charge-induced kinetic energy distortion ΔE_{SC} (i.e., spectral broadening and energy shift of the same order, see Saule et al. [18], Methods) in state-of-the-art ultrafast PES experiments versus pulse repetition rate. Colors of the symbols: different photon energy ranges (see legend). For each experiment the spot diameter is given, where available. For comparison, the energy distortion (heat map) is calculated relating the number of released PE to a 50-μm-diameter spot (typical size). For photon energies >40 eV (typical for attosecond-PES), a space-charge-limited spectral resolution of 0.1 eV corresponds to a number of ~10^4 photoelectrons released from the sample for each XUV pulse, resulting in an admissible number of ~1.5×10^5 impinging photons (see Saule et al. [18], Methods). Note that the photoelectron flux in Refs. [20, 24, 25, 31, 38] is generated by single-harmonic XUV excitation. Adapted with permission from [18] © Springer Nature

Thus, at a space-charge-limited photoelectron flux and at repetition rates in the kHz range, typical for contemporary attosecond-PES experiments and which imply temporal duty cycles of PE detection of a fraction of a percent [18], recording (multi-dimensional) PE spectra from solids with sufficient signal-to-noise ratio necessitates impractically—or even prohibitively—long acquisition times of several hours [19] to several tens of hours [17]. Over such periods, apart from the inconveniently long duration of measurement campaigns, laser instabilities and sample contamination constitute severe technological challenges.

Beyond time-resolved PES, other attosecond spectroscopies would profit from the availability of HHG sources with greatly improved repetition rates. In fact, in any technique involving the measurement of a *linear* sample absorption of HHG radiation, increasing the pulse repetition rate represents an immediate means of increasing the signal-to-noise ratio owing to improved statistics and to the possibility to use techniques such as lock-in amplification at modulation frequencies beyond the acoustic spectrum. In addition, providing the high peak intensities necessary for HHG at multi-MHz repetition rates would improve the disentanglement of the sought-for single-particle response from macroscopic phase-matching effects in HHG spectroscopy.

Last but not least, frequency-comb-driven HHG at multi-MHz pulse repetition rates transfers the comb properties of the driving pulse train [40] to the extreme-ultraviolet spectral region [41, 42], promising high-resolution spectral measurements [43, 44].

In conclusion, the successful extension of attosecond metrology to pulse repetition rates of several (tens of) MHz, permitting to fully tap the potential of its various techniques for science and technology, has provided a strong and urgent impetus for the development of a novel generation of HHG sources, combining *high output photon energies*, *high photon flux* and *high repetition rates*. The research activities summarized in this book addressed the development of such a source.

Section 1.2 introduces PES-based attosecond metrology, the main application having motivated the research reported here, in more detail, and femtosecond enhancement cavities, our technology of choice for tackling the above-mentioned challenge. Section 1.3 starts with a retrospective on this field preceding this work, from which the research objectives and the structure of this work are derived.

1.2 Photoemission-Spectroscopy-Based Attosecond Metrology

1.2.1 Photoelectron Spectroscopy—A Brief Historical Overview

Particularly direct access to electronic structure and dynamics is afforded by photoemission spectroscopy (PES) techniques [45]. In PES, a beam of photons with well-known properties—such as energy distribution, temporal structure, polarization, intensity and geometry—impinges on the investigated sample, releasing bound electrons via the photoelectric effect. A photoelectron (PE) detector records their kinetic energy, often along with momentum (ARPES: angle-resolved PES) or with spatial (PEEM: PE emission microscopy) resolution. The probability for a photon to release an electron with a certain momentum and energy depends on the initial, intermediate, and final quantum states that participate in photoemission. Thus, together with knowledge of the properties of the impinging light, PE spectra provide the perhaps most direct information about these states.

With Hertz discovery that the irradiation of electrodes with ultraviolet light facilitates the creation of an electric arc [20] in 1887, followed by a series of more refined experiments leading to Einstein's explanation of the photoelectric effect [21] in 1905, the first PES experiments date back more than a century and were instrumental for the development of quantum mechanics. However, it was not until about half a century later that the technological prerequisites were made available, that allowed PES to start a rapid evolution toward a highly versatile, widespread and increasingly powerful analytical tool for fields as diverse as material and surface science, chemical analysis and nanotechnology.

Two essential milestones along this road were, on the one hand, progress in high-vacuum technology and, on the other hand, the advent of synchrotron radiation (SR) sources. The former, besides ensuring a collision-free travel of PE to the detector, constituted a necessary condition for reliable and reproducible experiments at solid surfaces because of the extreme sensitivity of PES to their atomic configuration. Nowadays, (ultra-)high-vacuum systems with pressures on the order of 10^{-10} mbar are a matter of course in scientific laboratories, ensuring the maintenance of atomically clean surfaces of solids for hours.[4] Furthermore, SR sources have afforded high-brightness, monochromatic radiation with photon energies tunable across the (extreme) ultraviolet (5–100 eV), soft X-ray (100–1000 eV) and hard X-ray (>1000 eV) spectral regions, thus suitable to efficiently liberate electrons both

---

[4] Assuming a sticking coefficient of 1 (i.e., that each particle hitting the surface under test adheres to it), at a pressure of 10^{-10} mbar, the buildup of a monolayer on an atomically clean surface of a solid takes ~2 h.

from the valence levels as well as from the more tightly bound core levels via single-photon photoemission. Accompanied by progress of PE detector technology, high-resolution[5] PES carried out with monochromatic SR has evolved to an indispensable metrology tool for studies of the electronic energy levels of atoms and molecules, as well as for mapping the full band structure of solids. These measurements provide a *static* picture of the electronic configuration of the sample under scrutiny.

The advent of high-energy, ultrashort-pulsed lasers has constituted another milestone in PES, endowing this metrology with temporal resolution on the native scale of (correlated) electron dynamics in matter, in multiple ways. At large-scale facilities for example [22], femtosecond laser pulses have been employed to modulate the energy of electrons within a bunch with a typical duration of a few tens of picoseconds. By subsequent spatial separation of the energy-modulated electrons, this "laser slicing" method [23] permits temporal resolutions on the order of 100 fs (sufficient to access atomic vibrational dynamics), and has been in operation at several synchrotron facilities. Furthermore, free-electron lasers (FEL) operated in the self-amplified spontaneous emission (SASE) mode afford high-photon-energy pulses concentrated in spikes with a temporal envelope between several tens and a few hundred femtoseconds (following that of their parent electron bunch), and currently, considerable efforts address the challenge of temporally controlling and confining their emission to sub-fs durations—with the help of optical ultrashort-pulsed lasers [24].

Another major contribution of femtosecond lasers to PES has been the ability to perform measurements with femtosecond temporal resolution in table-top implementations. In the 1980s, pump-probe experiments in the *two-photon* PES configuration have enabled first direct, time-domain studies of surface-state dynamics with sub-100-fs temporal resolution [25]. Here, the sample is excited by an ultrashort pump laser pulse, and the subsequent evolution of the sample is followed by a second probe pulse, releasing photoelectrons from a surface state to vacuum and, thus, providing "snapshots" of the dynamically evolving material system at well-defined temporal delays with respect to the pump pulse. The coherent excitation of several quantum states, together with PES probing confined to a (variably delayed) time window considerably shorter than the lifetimes of these states, thus revealed the relative *phases* of the wavefunctions of surface states in addition to their magnitudes [26].

1.2.2 Attosecond Metrology Based on Photoelectron Spectroscopy

In its first two decades, *attosecond-PES employing HHG* has studied physical phenomena that had been hidden to previous experimental tools. Prominent examples employing isolated attosecond pulses and the *attosecond streaking* [27, 28] technique, include measurements of delays in photoemission from different orbitals with

---

[5] The resolving power in a PES measurement is given by the ratio $\Delta E / E$ so that from this point of view, the use of photon energies as low as possible benefits energy resolution.

attosecond [29] and, most recently, even sub-attosecond temporal resolution [46] or the real-time observation of light-induced electron tunneling [47]. Furthermore, the complementary *RABBITT* [47, 48] technique (reconstruction of attosecond harmonic beating by interference of two-photon transitions), using the trains of attosecond pulses generated by driving HHG with multi-cycle optical fields, enables a unique combination of high temporal resolution (owing to the brevity of the individual attosecond pulses in the train) with narrowband spectral excitation stemming from spectrally isolated harmonics rather than a continuous spectrum [19]. Very recently, the particular potential of this combination for the investigation of electron dynamics in condensed matter has been demonstrated with angle-resolved RABBITT measurements on metals, revealing differences in lifetimes between photoelectrons born into free-electron-like states and those excited into unoccupied excited states [17], and allowing the distinction between fundamental electron–electron interactions, such as scattering and screening [49].

Besides widening the frontiers of basic science, the extension of the measurement and control of time-varying processes to the characteristic timescale of dynamics in the electronic shells of atoms holds the promise of novel, groundbreaking technologies. Among those, signal processing with optical fields at PHz switching rates and the development of novel material systems for efficiently harvesting sunlight constitute just a few examples.

Yet, so far, the majority of attosecond-PES experiments has been carried out on gas-phase samples, with the extension of these experimental techniques to solids facing serious technological challenges. A particularly severe shortcoming of laser systems employed in state-of-the-art attosecond PES experiments arises from their relatively low pulse repetition rates, in the context of space charge, as discussed in Sect. 1.1.

Among state-of-the-art HHG sources, *femtosecond enhancement cavities* (EC) currently constitute the laser architecture most successful at combining high photon energies, high photon flux and high repetition rates (Fig. 1.1). In EC, the unconverted portion of the energy of the driving visible/near-infrared (VIS/NIR) pulses is "recycled" after each interaction with the gas target, uniquely affording high peak intensities necessary for efficient HHG in large focal volumes, despite the moderate pulse energy available from multi-MHz-repetition-rate lasers. Since their first demonstrations [41, 42] in 2005, concentrated development efforts have advanced this technology to a maturity that renders cavity-enhanced HHG sources suitable both for precision measurements for quantum transitions [43, 44, 50] and for space-charge-free (attosecond-temporal-resolution) PES of solids [18, 51].

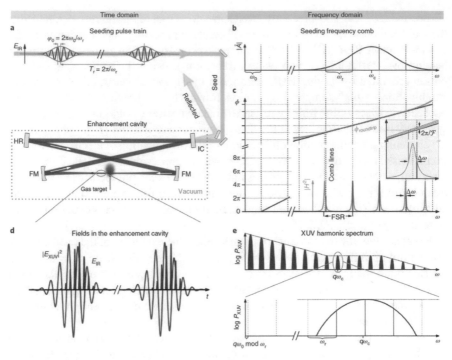

Fig. 1.2 Mode of operation of cavity-enhanced HHG. **a** A train of pulses with an envelope repetition period $T_r = 2\pi/\omega_r$ and a pulse-to-pulse carrier-envelope offset slippage $\varphi_0 = 2\pi\omega_0/\omega_r$, where ω_0 and ω_r are the frequency-comb parameters (see text), is coupled to a passive resonator. IC: partially transmitting input coupling mirror, HR: highly reflective mirror, FM: focusing HR. The choice of small angles of incidence on the resonator mirrors (as typical for dielectric optics) together with beam folding aiming at a compact footprint, typically result in a bowtie geometry for the ring resonator. **b, c** Frequency-domain representation of the spectra of the input pulse train (**b**) and of the squared magnitude of the resonator transfer function (**c**). Unlike the frequency comb modes, the cavity resonances are not equidistant in the frequency domain, which is a consequence of residual dispersion of the roundtrip phase $\phi_{roundtrip}(\omega)$ (shown in blue). Thus, the free-spectral range (FSR) is ω-dependent. **d** Inserting a gas target at a focus, see (**a**), allows for the generation of XUV pulse trains, emerging via HHG: each half-cycle of the driving field emits an XUV burst, rigidly locked to, and lagging the former by the electron excursion time, and whose intensity and energy spectrum depends on the field amplitude. **e** Generated XUV spectrum, consisting of individual harmonics, which emerge via spectral interference of the XUV bursts within the attosecond pulse train, cf. (**d**). The sub-structure of a harmonic, consists of an XUV frequency comb with the original repetition frequency ω_r and an offset frequency equal to $q\omega_0$, where q denotes the harmonic order. Adapted with permission from [52] © Springer Nature

1.2.3 Femtosecond Enhancement Cavities

The mode of operation of an EC[6] is illustrated in Fig. 1.2. The multi-MHz-repetition-rate VIS/NIR pulse train emitted by a (phase-stabilized) modelocked-oscillator-based

---

[6] This section proceeds in close analogy to the review manuscript [52].

laser system (Fig. 1.2a) is coupled to a free-space, passive optical resonator, the EC, by means of a partially transmitting input coupling mirror. The resonator is normally set up to fulfil stability, which ensures low-diffraction-loss propagation of a spatially coherent light beam within well-defined transverse eigenmodes [53].

Typically, a ring (i.e. traveling-wave) resonator is employed in order to limit the number of interactions with the intracavity nonlinear medium per roundtrip to one. Efficient coupling of the pulse train impinging on the input coupler can be achieved if the electric field of each input pulse constructively interferes with the pulse circulating in the resonator at the input coupling mirror. Under this resonance condition, and if the roundtrip losses in the resonator are sufficiently low, the energy of the circulating pulse can exceed that of the individual input pulses by a few orders of magnitude. This energy enhancement enables intracavity focal regions with peak intensities on the order of 10^{14} W/cm^2 and beam waists of a few 10s of μm, affording efficient HHG in a high-pressure gas target at repetition rates as high as several tens of MHz.

Figure 1.2b, c show the frequency-domain representation of the excitation of a passive optical resonator with a frequency comb. The (phase-stabilized) train of temporally equidistant input pulses (Fig. 1.2a) exhibits a spectral structure consisting in a multitude of equidistant narrow lines, forming an optical frequency comb [40, 54] (Fig. 1.2b). The optical (angular) frequency ω_N of each such "comb mode" can be parametrized as $\omega_N = \omega_0 + N\omega_r$, where $\omega_r = 2\pi/T_r$ and ω_0 are two radio frequencies, corresponding to the inverse roundtrip time T_r of the optical pulse inside the modelocked oscillator, and to an overall comb offset frequency, respectively, and N is an integer.

The resonance condition for optimum energy enhancement requires that both the repetition period and the pulse-to-pulse phase slippage of the optical carrier wave underneath the field envelope traveling in the EC be matched to the corresponding parameters of the input pulse train, i.e., $T_r = 2\pi/\omega_r$ and $\varphi_0 = 2\pi\,\omega_0/\omega_r$, respectively. In the frequency domain, this condition is equivalent to optimizing the overlap of the input frequency comb with the resonance spectrum of the EC. The latter is the squared magnitude of the resonator transfer function $H(\omega)$, that for a lossless input coupler with reflectivity $R(\omega)$, for a cavity roundtrip attenuation $A(\omega)$ and a roundtrip phase $\phi_{roundtrip}(\omega)$ reads:

$$H(\omega) = \frac{\sqrt{1 - R(\omega)}}{1 - \sqrt{R(\omega)A(\omega)}e^{i\phi_{roundtrip}(\omega)}}. \tag{1.1}$$

The frequency-dependent power enhancement $|H(\omega)|^2$ (Fig. 1.2c) exhibits a comb-like structure of sharp resonances. In contrast to the spectrum of the frequency comb (cf. Fig. 1.2b), the resonances of a linear EC are usually not equidistant due to residual chromatic dispersion [30] (Fig. 1.2c). This brings about a spectral filtering, limiting the bandwidth that can be simultaneously enhanced and, therefore, the pulse duration achievable in linear ECs. This effect increases for narrower resonances, i.e., for higher resonator finesse $\mathcal{F}(\omega) \approx 2\pi/[1 - R(\omega)A(\omega)]$. As an orientation [31], the power enhancement drops in good approximation to half its peak value if the roundtrip phase $\phi_{roundtrip}(\omega)$ deviates from a linear phase by $\pi/\mathcal{F}(\omega)$ (Fig. 1.2c).

The importance of precisely controlling the roundtrip phase triggered the development of several sensitive spectral-phase measurement techniques [30, 31, 55–57]. Furthermore, while the roundtrip period of a pulse circulating in the EC can in good approximation be adjusted via the physical cavity length, for a given spectral coverage of the input frequency comb, there is a well-defined value for $\varphi_0 = 2\pi\omega_0/\omega_r$ that maximises the power enhancement in a certain spectral range [58]. This is due to several processes affecting the carrier-envelope phase of the pulse upon propagation along the optic axis of the resonator, including reflections off multilayer mirrors, Gouy phase shifts accumulated upon propagation through foci, or transmission through any dispersive elements along the beam path. The corresponding optimum comb-offset frequency ω_0 is the intercept of the linear fit to the roundtrip phase (red line) with the abscissa in Fig. 1.2c.

For most applications, the offset frequency of the seeding comb can be adjusted to optimise broadband enhancement. However, particular applications, such as the enhancement of waveform-stable pulses for the generation of isolated attosecond pulses via intracavity gating mechanisms [32] require careful tuning of the offset frequency. Apart from (slight) tuning by intracavity dispersion or Gouy phase shifts, it has been shown that this can be achieved over a wide range by means of multilayer mirror design [58].

The inclusion of a nonlinear medium, such as a gas target, in a high-intensity (focal) region of an EC introduces nonlinear dynamics, leading to a considerable deviation from the linear-EC response (see Sect. 2.2). However, for finesse values in the range of several hundred to a thousand, power enhancement factors of 35–300 are routinely reached nowadays in the presence of intracavity high-pressure gas targets and the periodicity of these nonlinearities with ω_r preserves the frequency-comb structure of the intracavity light. Typically, the circulating pulses comprise several oscillation cycles, resulting in a train of XUV attosecond pulses per driving pulse, rigidly locked to the field of the latter (Fig. 1.2d).

Figure 1.2e illustrates the emergence of the XUV frequency comb. The generated XUV spectrum consists of high harmonics, of odd order of the driving spectrum [8] (upper panel) and extending over several tens of eV. The sub-structure of each harmonic (lower panel) is that of a frequency comb with a repetition frequency equal to that of the original comb and an offset frequency equal to $q\omega_0$, where q denotes the harmonic order [41, 42].

1.2.4 Experimental Implementations and Typical Parameters for Cavity-Enhanced HHG

The first ECs for HHG were seeded by frequency combs delivered by Ti:Sa oscillators [41, 42, 59], emitting spectra centred at approximately 800 nm. The average powers typical to these systems, on the order of 1 W, required high-finesse cavities in order to provide intracavity focal regions with peak intensities of 10^{13} W/cm^2 or more, as necessary for HHG. With (spectrally-averaged) finesse values of up to 2000 [60], these ECs supported circulating average powers between a few tens [41] and a few hundred [42, 59] Watts.

The short pulse durations typical to Ti:Sa oscillators, together with the aforementioned spectral filtering (particularly pronounced at high finesse) led to a roughly 1.5-fold temporal elongation of the circulating pulses with respect to the seeding pulses, with the shortest reported circulating pulse duration of 28 fs [41]. A Ti:Sa-based frontend employing an injection-locked femtosecond amplification cavity provided seeding pulses with higher average power (6 W) and narrower spectrum (supporting 80-fs duration), which resulted in 1-kW-level circulating pulses without temporal elongation [61].

Shortly after the demonstration of the first Ti:Sa-based EC-HHG systems, the advent of high-power Yb-based master-oscillator-power-amplifier lasers afforded multi-MHz seeding pulse trains with average powers on the order of 100 W and pulse durations in the 100–250 fs range, spectrally centred around 1030 nm (see also next section). With these seeding lasers, multi-kW-range intracavity average powers became a matter of course [18, 43, 50, 51, 62–67]—at moderate resonator finesse. The highest average powers reported to date for an empty EC are 670 kW and 400 kW for a 250-MHz train of 10-ps and 250-fs pulses, respectively [68], employing resonator designs with large illumination spots on all optics [33].

In addition, nonlinear spectral broadening and temporal compression of the output of high-power Yb-based lasers provided a convenient means of decreasing the pulse duration to the few-10-fs regime [63]. In empty ECs, for 30-fs pulses, 20 kW of average power have been demonstrated [67] and the shortest pulse reported in a kW-level EC had a Fourier-transform limit of 17.5 fs, corresponding to 5.4 oscillation cycles of the carrier electric field [31].

The pulse repetition rates of EC-HHG systems reported to date range from 10 MHz [59, 65] to 250 MHz [67]. Lower repetition rates are impeded by large resonator lengths. At higher repetition rates, the efficiency of HHG suffers from cumulative plasma effects (see Sect. 2.5).

1.3 Research Objectives and Structure of the Book

1.3.1 The Project MEGAS (MHz Attosecond Pulses for Photoelectron Spectroscopy and Microscopy)

In the remainder, this book summarizes the development of femtosecond enhancement cavities as an enabling technology for next-generation attosecond metrology, led by the author between the years 2013 and 2019, at the *Laboratory of Attosecond and High-Field Physics* (LAP), jointly located at the Max Planck Institute for

Quantum Optics (MPQ) and the Ludwig Maximilian University Munich (LMU) and headed by Prof. Ferenc Krausz.

In the early 2000s, at the MPQ and LMU, cavity-enhanced HHG had been initiated jointly at LAP and in the *Laser Spectroscopy* division headed by Prof. T. W. Hänsch, with the common goal to build sources of coherent vacuum/extreme-ultraviolet (in the following, for simplicity referred to as XUV) radiation with properties akin to those of (phase-stabilized) modelocked lasers (i.e., frequency combs). In particular, such a source should generate a multi-MHz-repetition-rate train of broadband XUV pulses, rigidly locked to the controlled electric field of the driving visible/near-infrared [41, 59] pulses.

An immediate motivation for frequency-domain metrology has been the prospect to measure narrow electronic transitions employing frequency-comb techniques [40, 54], such as the 1S–2S transition in singly-ionized helium at 61 nm, for precision tests of bound-state quantum electrodynamics [34]. For time-domain measurements pursued at LAP, the severe experimental constraints introduced by space-charge effects in laser-assisted attosecond-PES at repetition rates $\gg$ 1 MHz constituted a decisive impetus for these developments (see Sect. 1.1). Furthermore, passive optical resonators had been demonstrated to boost the overall efficiency of nonlinear conversion processes [35, 69]. Given the low conversion efficiency of HHG (typically $\ll$ 10^{-5}), ECs promised a route toward high-power, broadband, compact XUV sources with a brilliance and a pulse repetition rate similar to those of synchrotron radiation [45].

Around 2005, Yb-based ultrafast laser systems were emerging as an average-power-scalable alternative to the well-established Ti:Sa technology [3], delivering pulses with durations around 200 fs and a—at that time—new combination of high average powers (of several tens of Watts) and high repetition rates (of several tens of MHz). At LAP, an Yb-fiber-amplifier laser system [70] allowed us to address three key questions/challenges, the results of which would constitute the basis for the research program that eventually led to the development of the first multi-MHz-repetition-rate high-XUV-energy attosecond-PES beamline:

i.   The aforementioned combination of high input power and high repetition rate allowed us to demonstrate intracavity ultrashort pulses with several tens of kW of average power [62] (2-ps pulses with 72 kW and 200-fs pulses with 18 kW). On the one hand, this constituted a new power regime for ultrashort pulses. Together with the possibility of further reducing the seeding pulse duration via nonlinear temporal compression in solid-core fibers [70], this promised unprecedented conditions for driving the low-efficiency nonlinear process of HHG. On the other hand, these experiments revealed intensity-related damage limitations of the cavity optics, prompting in-depth studies of advanced cavity designs in the subsequent years.

ii. Any linear and nonlinear temporal distortion of the pulse upon a roundtrip in an EC is enhanced by the latter. This includes (unavoidable) higher-order phase distortions introduced by reflections off the multi-layer dielectric mirrors and nonlinear phase acquired upon interaction with intracavity nonlinear media, and imposes stringent conditions to the roundtrip dispersion. To measure the latter, we applied broadband spatial-spectral interferometry to EC for the first time [57], devising a metrology that would later also prove useful for investigations of nonlinear EC [71].

iii. One of the main challenges en route to validating the EC technology for attosecond metrology was coupling out the intracavity generated harmonics for their use outside of the cavity. In contrast to previously demonstrated reflection-[41, 42] or diffraction-based [36] XUV output coupling techniques, our group pursued the idea of geometrically coupling out the harmonics through on-axis hard apertures in a cavity mirror [37]. Geometric output coupling ensures low distortions to the participating pulses as well as a solution for high output coupling efficiencies for high photon energies. Our initial studies in this direction addressed tailoring the spatial mode of a EC [38], and triggered multiple approaches, ideas and results in the subsequent years.

The milestones (i)–(iii), together with first cavity-enhanced HHG demonstrations with kW-class EC around 2011/2012 [43, 61–63] rendered EC the most promising route toward realizing an HHG source combining high photon energies, a high photon flux and a high repetition rate for the first time. The project MEGAS—*MHz Attosecond Pulses for Photoelectron Spectroscopy and Microscopy* brought together expertise from the Fraunhofer Institutes for Applied Optics and Precision Engineering (IOF) in Jena and for Laser Technology (ILT) in Aachen as well as from the MPQ and LMU in Garching/München with the goal of building and validating such a source, between 2014 and 2018.

Currently, this beamline is in operation for attosecond-time-resolution angle-resolved photoelectron spectroscopy at the MPQ. First experiments attest a reduction of the measurement time by a factor between 100 and 1000 over state-of-the-art attosecond-PES experiments performed under identical space-charge conditions [18, 39], reducing the measurement time from several days to minutes.

1.3.2 Structure of the Book

Chapter 2 summarizes our systematic efforts towards setting up the MEGAS beamline, carried out at several experimental setups. The results are arranged thematically rather than chronologically. The first three sections contain the natural continuations of the directions opened up by the milestones (i)–(iii) described in the previous section:

- Section 2.1 addresses **the power scaling** of EC's by means of advanced designs of the cavity geometry and optics, enabling the enhancement of pulses as short as a few tens of femtoseconds to unprecedented average power levels in the multi-kW-level range. This constituted the first "ingredient" for boosting the XUV photon flux and energies of cavity-enhanced HHG with respect to the state of the art prior to this work. In addition, our studies on high-power resonator designs led to the demonstration of intracavity picosecond pulses with average powers approaching 1 MW, which are likely to benefit other applications, such as hard X-ray generation via Thomson scattering off relativistic electrons.

- Section 2.2 summarizes our **quantitative study of the self-phase-modulation-related intensity clamping** observed in cavity-enhanced HHG, which allowed for the derivation of an optimization routine for the main design parameters of the HHG process. This study was instrumental for the ability to make use of the newly available power regime for an improved HHG conversion efficiency.
- Section 2.3 reviews our work on **geometric XUV output coupling**. As a last, crucial ingredient for an EC-based XUV beamline, the high-harmonic radiation had to be efficiently coupled out of the high-power EC and with low distortions. We identified, thoroughly studied and optimized geometric output coupling of the intracavity generated XUV light through on-axis hard apertures in the EC mirror following the HHG focus, demonstrating output coupling of harmonics with photon energies > 100 eV with efficiencies on the order of 10% or higher.
- Section 2.4 describes the **MEGAS beamline**, including the phase-stabilized, repetition-rate tunable femtosecond frontend and the enhancement cavity. Studies on the cumulative plasma effects in the HHG target are presented, which allowed for an optimization of the repetition rate of the system.
- Section 2.5 presents **PES results obtained with the MEGAS system**, in particular the first attosecond-temporal resolution, space-charge-free, angle-resolved PES measurements at photon energies approaching 100 eV and at a repetition rate >> 1 MHz and at a temporal duty cycle of photoelectron detection close to 100%.

Chapter 3 gives an outlook of possible further developments of this technology:

- Section 3.1 discusses **prerequisites for the generation of isolated attosecond pulses with EC**. These include our experimental studies toward shortening the duration of the pulse(s) circulating in the EC while maintaining a power enhancement justifying the technological effort. The passive enhancement of pulses as short as 5.5 cycles of light with kW-level average power is demonstrated. Furthermore, this section includes our experiments on the enhancement of waveform-stable pulses. In Sect. 3.2, our comprehensive theoretical and experimental study on temporal gating methods suitable for the generation of isolated attosecond pulses with EC is summarized. Among those, first experimental steps toward the implementation of the "attosecond lighthouse" inside a EC via a new method ("transverse mode gating") are presented.
- Section 3.3 addresses the **first demonstration of temporal dissipative solitons in a free-space EC**, which arose from our intensive preoccupation with passive optical resonators in the nonlinear regime. This novel manifestation of ultra-short pulses is characterized by a highly interesting combination of properties: self-stabilization, temporal self-compression, high peak power enhancement and excellent spatial and temporal profile of the beam. Besides an application in HHG, this novel regime of EC might provide opportunities for other nonlinear conversion processes, such as optical rectification, or for various linear and nonlinear spectroscopies.

References

1. H.A. Haus, Mode-locking of lasers. IEEE J. Sel. Top. Quantum Electron. **6**, 1173–1185 (2000)
2. A.H. Zewail, Femtochemistry: atomic-scale dynamics of the chemical bond. J. Phys. Chem. A **104**, 5660–5694 (2000)
3. T. Brabec, F. Krausz, Intense few-cycle laser fields: frontiers of nonlinear optics. Rev. Mod. Phys. **72**, 545–591 (2000)
4. F. Krausz, M. Ivanov, Attosecond physics. Rev. Mod. Phys. **81**, 163–234 (2009)
5. R. Geneaux, H.J.B. Marroux, A. Guggenmos, D.M. Neumark, S.R. Leone, Transient absorption spectroscopy using high harmonic generation: a review of ultrafast X-ray dynamics in molecules and solids. Philos. Trans. R. Soc. A **377**, 20170463 (2019)
6. J.L. Krause, K.J. Schafer, K.C. Kulander, High-order harmonic generation from atoms and ions in the high intensity regime. Phys. Rev. Lett. **68**, 3535–3538 (1992)
7. P.B. Corkum, Plasma perspective on strong field multiphoton ionization. Phys. Rev. Lett. **71**, 1994–1997 (1993)
8. M. Lewenstein, P. Balcou, M.Y. Ivanov, A. L'Huillier, P.B. Corkum, Theory of high-harmonic generation by low-frequency laser fields. Phys. Rev. A **49**, 2117–2132 (1994)
9. T. Popmintchev, M.-C. Chen, D. Popmintchev, P. Arpin, S. Brown, S. Alisauskas, G. Andriukaitis, T. Balciunas, O.D. Mucke, A. Pugzlys, A. Baltuska, B. Shim, S.E. Schrauth, A. Gaeta, C. Hernandez-Garcia, L. Plaja, A. Becker, A. Jaron-Becker, M.M. Murnane, H.C. Kapteyn, Bright coherent ultrahigh harmonics in the keV X-ray regime from mid-infrared femtosecond lasers. Science **336**, 1287–1291 (2012)
10. P.M. Kraus, M. Zürch, S.K. Cushing, D.M. Neumark, S.R. Leone, The ultrafast X-ray spectroscopic revolution in chemical dynamics. Nat. Rev. Chem. **2**, 82–94 (2018)
11. W.T. Pollard, R.A. Mathies, Analysis of femtosecond dynamic absorption spectra of nonstationary states. Annu. Rev. Phys. Chem. **43**, 497–523 (1992)
12. J.P. Marangos, Development of high harmonic generation spectroscopy of organic molecules and biomolecules. J. Phys. B At. Mol. Opt. Phys. **49**, 132001 (2016)
13. N. Dudovich, O. Smirnova, J. Levesque, Y. Mairesse, M.Y. Ivanov, D.M. Villeneuve, P.B. Corkum, Measuring and controlling the birth of attosecond XUV pulses. Nat. Phys. **2**, 781–786 (2006)
14. S. Hellmann, K. Rossnagel, M. Marczynski-Bühlow, L. Kipp, Vacuum space-charge effects in solid-state photoemission. Phys. Rev. B **79**, 035402 (2009)
15. N.M. Buckanie, J. Göhre, P. Zhou, D. von der Linde, M. Horn-von Hoegen, F.-J. Meyer zu Heringdorf, Space charge effects in photoemission electron microscopy using amplified femtosecond laser pulses. J. Phys. Condens. Matter **21**, 314003 (2009)
16. S.H. Chew, C. Späth, A. Wirth, J. Schmidt, S. Zherebtsov, A. Guggenmos, A. Oelsner, N. Weber, J. Kapaldo, A. Gliserin, M.I. Stockman, M.F. Kling, U. Kleineberg, Time-of-flight-photoelectron emission microscopy on plasmonic structures using attosecond extreme ultraviolet pulses. Appl. Phys. Lett. **100**, 051904 (2012)
17. Z. Tao, C. Chen, T. Szilvasi, M. Keller, M. Mavrikakis, H. Kapteyn, M. Murnane, Direct time-domain observation of attosecond final-state lifetimes in photoemission from solids. Science **353**, 62–67 (2016)
18. T. Saule, S. Heinrich, J. Schötz, N. Lilienfein, M. Högner, O. de Vries, M. Plötner, J. Weitenberg, D. Esser, J. Schulte, P. Russbueldt, J. Limpert, M.F. Kling, U. Kleineberg, I. Pupeza, High-flux ultrafast extreme-ultraviolet photoemission spectroscopy at 18.4 MHz pulse repetition rate. Nat. Commun. **10**, 458 (2019)
19. M. Isinger, R.J. Squibb, D. Busto, S. Zhong, A. Harth, D. Kroon, S. Nandi, C.L. Arnold, M. Miranda, J.M. Dahlström, E. Lindroth, R. Feifel, M. Gisselbrecht, A. L'Huillier, Photoionization in the time and frequency domain. Science **358**, 893–896 (2017)
20. H. Hertz, Ueber einen Einfluss des ultravioletten Lichtes auf die electrische Entladung. Ann. Phys. Chem. **267**, 983–1000 (1887)
21. A. Einstein, Über einen die Erzeugung und Verwandlung des Lichtes betreffenden heuristischen Gesichtspunkt. Ann. Phys. **322**, 132–148 (1905)

22. S. Yamamoto, I. Matsuda, Time-resolved photoelectron spectroscopies using synchrotron radiation: past, present, and future. J. Phys. Soc. Jpn. **82**, 021003 (2013)
23. R.W. Schoenlein, Generation of femtosecond pulses of synchrotron radiation. Science **287**, 2237–2240 (2000)
24. E.A. Seddon, J.A. Clarke, D.J. Dunning, C. Masciovecchio, C.J. Milne, F. Parmigiani, D. Rugg, J.C.H. Spence, N.R. Thompson, K. Ueda, S.M. Vinko, J.S. Wark, W. Wurth, Short-wavelength free-electron laser sources and science: a review. Rep. Prog. Phys. **80**, 115901 (2017)
25. J. Bokor, Ultrafast dynamics at semiconductor and metal surfaces. Science **246**, 1130–1134 (1989)
26. U. Höfer, Time-resolved coherent photoelectron spectroscopy of quantized electronic states on metal surfaces. Science **277**, 1480–1482 (1997)
27. J. Itatani, F. Quéré, G.L. Yudin, M.Y. Ivanov, F. Krausz, P.B. Corkum, Attosecond streak camera. Phys. Rev. Lett. **88** (2002)
28. R. Kienberger, E. Goulielmakis, M. Uiberacker, A. Baltuska, V. Yakovlev, F. Bammer, A. Scrinzi, T. Westerwalbesloh, U. Kleineberg, U. Heinzmann, M. Drescher, F. Krausz, Atomic transient recorder. Nature **427**, 817–821 (2004)
29. M. Schultze, M. Fieß, N. Karpowicz, J. Gagnon, M. Korbman, M. Hofstetter, S. Neppl, A.L. Cavalieri, Y. Komninos, T. Mercouris, C.A. Nicolaides, R. Pazourek, S. Nagele, J. Feist, J. Burgdörfer, A.M. Azzeer, R. Ernstorfer, R. Kienberger, U. Kleineberg, E. Goulielmakis, F. Krausz, V.S. Yakovlev, Delay in photoemission. Science **328**, 1658 (2010)
30. M.J. Thorpe, R.J. Jones, K.D. Moll, J. Ye, R. Lalezari, Precise measurements of optical cavity dispersion and mirror coating properties via femtosecond combs. Opt. Express **13**, 882 (2005)
31. N. Lilienfein, C. Hofer, S. Holzberger, C. Matzer, P. Zimmermann, M. Trubetskov, V. Pervak, I. Pupeza, Enhancement cavities for few-cycle pulses. Opt. Lett. **42**, 271 (2017)
32. M. Högner, T. Saule, S. Heinrich, N. Lilienfein, D. Esser, M. Trubetskov, V. Pervak, I. Pupeza, Cavity-enhanced noncollinear high-harmonic generation. Opt. Express **27**, 19675–19691 (2019)
33. H. Carstens, S. Holzberger, J. Kaster, J. Weitenberg, V. Pervak, A. Apolonski, E. Fill, F. Krausz, I. Pupeza, Large-mode enhancement cavities. Opt. Express **21**, 11606 (2013)
34. M. Herrmann, M. Haas, U.D. Jentschura, F. Kottmann, D. Leibfried, G. Saathoff, C. Gohle, A. Ozawa, V. Batteiger, S. Knünz, N. Kolachevsky, H.A. Schüssler, T.W. Hänsch, T. Udem, Feasibility of coherent XUV spectroscopy on the 1 S − 2 S transition in singly ionized helium. Phys. Rev. A **79** (2009)
35. R. Paschotta, P. Kürz, R. Henking, S. Schiller, J. Mlynek, 82% efficient continuous-wave frequency doubling of 106 μm with a monolithic MgO:LiNbO$_3$ resonator. Opt. Lett. **19**, 1325 (1994)
36. D.C. Yost, T.R. Schibli, J. Ye, Efficient output coupling of intracavity high harmonic generation. Opt. Lett. **33**, 1099 (2008)
37. K.D. Moll, R.J. Jones, J. Ye, Output coupling methods for cavity-based high-harmonic generation. Opt. Express **14**, 8189 (2006)
38. J. Weitenberg, P. Rußbüldt, T. Eidam, I. Pupeza, Transverse mode tailoring in a quasi-imaging high-finesse femtosecond enhancement cavity. Opt. Express **19**, 9551 (2011)
39. S. Heinrich, T. Saule, M. Högner, Y. Cui, V.S. Yakovlev, I. Pupeza, U. Kleineberg, Attosecond intra-valence band dynamics and resonant-photoemission delays in W(110). Nat. Commun. **12**, 3404 (2021)
40. T. Udem, R. Holzwarth, T.W. Hänsch, Optical frequency metrology. Nature **416**, 5 (2002)
41. C. Gohle, T. Udem, M. Herrmann, J. Rauschenberger, R. Holzwarth, H.A. Schuessler, F. Krausz, T.W. Hänsch, A frequency comb in the extreme ultraviolet. Nature **436**, 234–237 (2005)
42. R.J. Jones, K.D. Moll, M.J. Thorpe, J. Ye, Phase-coherent frequency combs in the vacuum ultraviolet via high-harmonic generation inside a femtosecond enhancement cavity. Phys. Rev. Lett. **94** (2005)
43. A. Cingöz, D.C. Yost, T.K. Allison, A. Ruehl, M.E. Fermann, I. Hartl, J. Ye, Direct frequency comb spectroscopy in the extreme ultraviolet. Nature **482**, 68–71 (2012)

44. A. Ozawa, Y. Kobayashi, VUV frequency-comb spectroscopy of atomic xenon. Phys. Rev. A **87**, 022507 (2013)
45. S. Hüfner, *Photoelectron Spectroscopy: Principles and Applications*, 3rd edn. Advanced Texts in Physics (Springer, 2003)
46. M. Ossiander, F. Siegrist, V. Shirvanyan, R. Pazourek, A. Sommer, T. Latka, A. Guggenmos, S. Nagele, J. Feist, J. Burgdörfer, R. Kienberger, M. Schultze, Attosecond correlation dynamics. Nat. Phys. **13**, 280–285 (2016)
47. M. Uiberacker, T. Uphues, M. Schultze, A.J. Verhoef, V. Yakovlev, M.F. Kling, J. Rauschenberger, N.M. Kabachnik, H. Schröder, M. Lezius, K.L. Kompa, H.-G. Muller, M.J.J. Vrakking, S. Hendel, U. Kleineberg, U. Heinzmann, M. Drescher, F. Krausz, Attosecond real-time observation of electron tunnelling in atoms. Nature **446**, 627–632 (2007)
48. P.M. Paul, Observation of a train of attosecond pulses from high harmonic generation. Science **292**, 1689–1692 (2001)
49. C. Chen, Z. Tao, A. Carr, P. Matyba, T. Szilvási, S. Emmerich, M. Piecuch, M. Keller, D. Zusin, S. Eich, M. Rollinger, W. You, S. Mathias, U. Thumm, M. Mavrikakis, M. Aeschlimann, P.M. Oppeneer, H. Kapteyn, M. Murnane, Distinguishing attosecond electron–electron scattering and screening in transition metals. Proc. Natl. Acad. Sci. **114**, E5300–E5307 (2017)
50. G. Porat, C.M. Heyl, S.B. Schoun, C. Benko, N. Dörre, K.L. Corwin, J. Ye, Phase-matched extreme-ultraviolet frequency-comb generation. Nat. Photonics **12**, 387–391 (2018)
51. C. Corder, P. Zhao, J. Bakalis, X. Li, M.D. Kershis, A.R. Muraca, M.G. White, T.K. Allison, Ultrafast extreme ultraviolet photoemission without space charge. Struct. Dyn. **5**, 054301 (2018)
52. I. Pupeza, C. Zhang, M. Högner, J. Ye, Extreme-ultraviolet frequency combs for precision metrology and attosecond science. Nat. Photonics **15**, 175–186 (2021)
53. A.E. Siegman, *Lasers* (University Science Books, 1986)
54. J. Ye, S.T. Cundiff, *Femtosecond Optical Frequency Comb: Principle, Operation, and Applications* (Springer, 2005)
55. A. Schliesser, C. Gohle, T. Udem, T.W. Hänsch, Complete characterization of a broadband high-finesse cavity using an optical frequency comb. Opt. Express **14**, 5975 (2006)
56. T.J. Hammond, A.K. Mills, D.J. Jones, Simple method to determine dispersion of high-finesse optical cavities. Opt. Express **17**, 8998 (2009)
57. I. Pupeza, X. Gu, E. Fill, T. Eidam, J. Limpert, A. Tünnermann, F. Krausz, T. Udem, Highly sensitive dispersion measurement of a high-power passive optical resonator using spatial-spectral interferometry. Opt. Express **18**, 26184 (2010)
58. S. Holzberger, N. Lilienfein, M. Trubetskov, H. Carstens, F. Lücking, V. Pervak, F. Krausz, I. Pupeza, Enhancement cavities for zero-offset-frequency pulse trains. Opt. Lett. **40**, 2165 (2015)
59. A. Ozawa, J. Rauschenberger, C. Gohle, M. Herrmann, D.R. Walker, V. Pervak, A. Fernandez, R. Graf, A. Apolonski, R. Holzwarth, F. Krausz, T.W. Hänsch, T. Udem, High harmonic frequency combs for high resolution spectroscopy. Phys. Rev. Lett. **100** (2008)
60. A.K. Mills, T.J. Hammond, M.H.C. Lam, D.J. Jones, XUV frequency combs via femtosecond enhancement cavities. J. Phys. B At. Mol. Opt. Phys. **45**, 142001 (2012)
61. J. Lee, D.R. Carlson, R.J. Jones, Optimizing intracavity high harmonic generation for XUV fs frequency combs. Opt. Express **19**, 23315 (2011)
62. I. Pupeza, T. Eidam, J. Rauschenberger, B. Bernhardt, A. Ozawa, E. Fill, A. Apolonski, T. Udem, J. Limpert, Z.A. Alahmed, A.M. Azzeer, A. Tünnermann, T.W. Hänsch, F. Krausz, Power scaling of a high-repetition-rate enhancement cavity. Opt. Lett. **35**, 2052 (2010)
63. I. Pupeza, S. Holzberger, T. Eidam, H. Carstens, D. Esser, J. Weitenberg, P. Rußbüldt, J. Rauschenberger, J. Limpert, T. Udem, A. Tünnermann, T.W. Hänsch, A. Apolonski, F. Krausz, E. Fill, Compact high-repetition-rate source of coherent 100 eV radiation. Nat. Photonics **7**, 608–612 (2013)
64. C. Benko, T.K. Allison, A. Cingöz, L. Hua, F. Labaye, D.C. Yost, J. Ye, Extreme ultraviolet radiation with coherence time greater than 1 s. Nat. Photonics **8**, 530–536 (2014)
65. A. Ozawa, Z. Zhao, M. Kuwata-Gonokami, Y. Kobayashi, High average power coherent VUV generation at 10 MHz repetition frequency by intracavity high harmonic generation. Opt. Express **23**, 15107 (2015)

66. A.K. Mills, S. Zhdanovich, A. Sheyerman, G. Levy, A. Damascelli, D.J. Jones, *An XUV Source Using a Femtosecond Enhancement Cavity for Photoemission Spectroscopy*, ed. by S.G. Biedron (2015), p. 95121I
67. H. Carstens, M. Högner, T. Saule, S. Holzberger, N. Lilienfein, A. Guggenmos, C. Jocher, T. Eidam, D. Esser, V. Tosa, V. Pervak, J. Limpert, A. Tünnermann, U. Kleineberg, F. Krausz, I. Pupeza, High-harmonic generation at 250 MHz with photon energies exceeding 100 eV. Optica **3**, 366 (2016)
68. H. Carstens, N. Lilienfein, S. Holzberger, C. Jocher, T. Eidam, J. Limpert, A. Tünnermann, J. Weitenberg, D.C. Yost, A. Alghamdi, Z. Alahmed, A. Azzeer, A. Apolonski, E. Fill, F. Krausz, I. Pupeza, Megawatt-scale average-power ultrashort pulses in an enhancement cavity. Opt. Lett. **39**, 2595 (2014)
69. A. Ashkin, G. Boyd, J. Dziedzic, Resonant optical second harmonic generation and mixing. IEEE J. Quantum Electron. **2**, 109–124 (1966)
70. T. Eidam, F. Röser, O. Schmidt, J. Limpert, A. Tünnermann, 57 W, 27 fs pulses from a fiber laser system using nonlinear compression. Appl. Phys. B **92**, 9–12 (2008)
71. S. Holzberger, N. Lilienfein, H. Carstens, T. Saule, M. Högner, F. Lücking, M. Trubetskov, V. Pervak, T. Eidam, J. Limpert, A. Tünnermann, E. Fill, F. Krausz, I. Pupeza, Femtosecond enhancement cavities in the nonlinear regime. Phys. Rev. Lett. **115** (2015)

Chapter 2
Cavity-Enhanced High-Order Harmonic Generation for Attosecond Metrology

2.1 Power Scaling of Femtosecond Enhancement Cavities

A crucial development in the first decade of this century, which was to boost EC technology shortly thereafter, was the advent of high-power Yb-doped fiber laser (chirped-pulse) amplifiers. These systems provided amplification of the output of 1-μm modelocked oscillators at their original repetition rates of several tens of MHz to trains of pulses with durations around 200 fs, and average powers in the 100-W range. The direct investigation of the power scaling limitations of standard-design ECs with such a system led to the insight that EC mirror damage was primarily caused by high intensity rather than high average power [1]. This result triggered a systematic investigation, both in theory and in experiment, of advanced resonator designs conceived to: (i) exhibit large spot sizes on *all* mirrors in order to mitigate intensity-induced damage and (ii) minimize the sensitivity of the mirror arrangement to misalignment.

2.1.1 Large-Mode Enhancement Cavities [2]

A theoretical study identified symmetric bow-tie ECs with a single tight focus (cf. [2]), operated close to their inner stability edge as the simplest, most appropriate design fulfilling the two above criteria. It is noteworthy that as the distance of the curved mirrors is decreased approaching the inner edge of stability (while keeping the total length of the ring resonator constant), the focus in the shorter path between the curved mirrors decreases in size and that in the longer path increases. Due to the latter property, operation "close to the inner stability edge" ensures large beam sizes on *all* resonator mirrors, with spot sizes varying among the mirrors by only a few percent in realistic implementations.

© The Author(s), under exclusive license to Springer Nature Switzerland AG 2022
I. Pupeza, *Passive Optical Resonators for Next-Generation Attosecond Metrology*,
SpringerBriefs in Physics, https://doi.org/10.1007/978-3-030-92972-5_2

Equally importantly, we investigated the sensitivity of different resonator designs to misalignments. To this end, we developed a metric evaluating the overlap integral of the fundamental resonator transverse eigenmode described by the complex transverse field distribution $\Psi_{initial}(x, y)$ with that after a geometric perturbation, $\Psi_{pert}(x, y)$:

$$U = \left| \int_{-\infty}^{\infty} \Psi_{initial}(x, y)\Psi^*_{pert}(x, y)dxdy \right|^2 . \tag{2.1}$$

The change of this overlap integral, $\Delta U = 1 - U$, can be applied to describe (and compare) the effect of a given misalignment for arbitrary resonator designs. Figure 2.1 shows an example for a perturbation consisting in tilting one plane mirror by an angle of 1 μrad.

In addition to identifying a preferred resonator design combining large spot sizes on all mirrors and robustness to misalignment, this theoretical study predicted that operation at the outer stability edge (i.e., curved mirrors further apart), as well as previously proposed symmetric resonator configurations exclusively consisting of curved mirrors [3] (all-curved-mirror cavity) would be highly sensitive to misalignment and, thus, impractical for implementation.

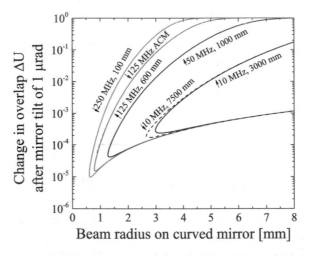

Fig. 2.1 Misalignment sensitivity of different passive optical resonator designs. The figure shows the transverse change ΔU of the fundamental resonator mode for a small tilt of one mirror (y axis), and the beam radius on the curved mirrors (x axis), see text for details, as a parametric plot of the distance between the two curved mirrors (i.e. position within the stability range). The numbers on the curves indicate the repetition frequency of the resonators (from which the length can be calculated) and the radius of curvature of the curved mirrors. ACM: all-curved-mirror cavity (see text). A small value of ΔU, and a large spot size on the mirrors are desirable. Adapted with permission from [2] © The Optical Society

In [2], the feasibility of actively stabilizing a resonator close to the inner stability edge with a power enhancement of 2000 and without any loss in stability was experimentally demonstrated at low powers, and the sensitivity to misalignment of the other designs was qualitatively confirmed in the experiment. In our proof-of-principle experiment, beam radii exceeding 5.5 mm ($1/e^2$-intensity) in the sagittal plane on all four mirrors of a standard bow-tie resonator (consisting of two plane and two spherical mirrors with a radius of curvature of 600 mm) with a roundtrip length of 1.2 m were demonstrated. Due to non-orthogonal incidence on the spherical mirrors, the beam size in the tangential plane was 2.6 mm. We proposed, however, means to avoid this ellipticity despite the above-mentioned non-zero angles of incidence on the curved mirrors.

These results constituted the basis for scaling femtosecond enhancement cavities well into the multi-kW average-power range.

2.1.2 Megawatt-Scale Average-Power Ultrashort Pulses in an Enhancement Cavity [4]

Employing a state-of-the-art Yb-fiber chirped-pulse amplification (CPA) laser system [5], we experimentally investigated the power scaling of large-mode enhancement cavities. The laser frontend emitted a 250-MHz repetition-rate train of 250-fs pulses with up to 500 W of average power, which could be either chirped to multi-picosecond durations by varying the grating distance on the CPA, or nonlinearly compressed to 30 fs, thus affording a broad parameter range of input pulses for the cavity.

We confirmed that the large spot sizes on the mirrors mitigated the primary cause of mirror damage (i.e. high intensity), allowing for record average powers of several 100 kW even for 250-fs pulses. However, at these power levels, absorption in the mirror coatings—even though in the few-ppm range—led to considerable thermal stress gradients in the mirror substrates, strongly affecting the transverse resonator mode (including the position in the stability range).

We quantitatively investigated these changes and derived a thermal sensitivity metric allowing for quantitative design and comparison of resonators with given mirror characteristics. These quantitative studies allowed us to devise enhancement cavities with world-record average powers and pulse energies in three different regimes (employing different mirrors), see also Fig. 2.2:

- For 10-ps pulses, we reached 670 kW of average power. This regime is particularly interesting for the generation of hard X-rays via Thomson scattering off relativistic electrons.
- For 250-fs pulses, we demonstrated 400 kW of average power. This regime is interesting for high-spectral-resolution RABBITT as well as XUV frequency-comb applications.

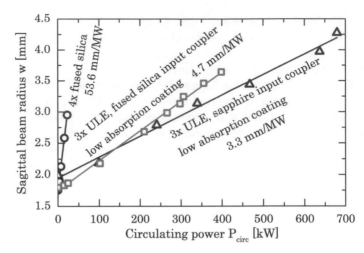

Fig. 2.2 Power scaling results with a 250-MHz train of pulses seeding an empty, large-mode enhancement cavity. The plot shows the evolution of the beam radius on the curved mirrors, as a function of the circulating average power, for different mirror substrate combinations. Due to the high thermal expansion coefficient of fused silica, an EC built exclusively with mirrors coated on fused-silica substrates shows the strongest thermal lensing effect. Using ultra-low expansion glass (ULE) for the high-reflectivity mirrors and a sapphire substrate for the input coupler led to the weakest thermal lensing effect and the highest achievable average power (close to 700 kW), with ~10-ps pulses. When decreasing the pulse duration, we observed damage in the input coupler, which we attributed to the modest optical quality of the sapphire substrate. Replacing the input coupler with one coated on a fused-silica substrate increased the thermal lensing effect, but enabled 250-fs pulses with 400 kW of average power. Adapted with permission from [4] © The Optical Society

- In a later publication [6], for 30-fs pulses we demonstrated 20 kW of average power, a regime paving the way to the first cavity-enhanced HHG experiments for attosecond metrology.

2.1.3 Balancing of Thermal Lenses in Enhancement Cavities with Transmissive Elements [7]

Besides allowing for a quantitative design optimization of high-power, large-mode cavities, the models developed in the previous two sections allowed us to investigate techniques for balancing the wavefront distortions introduced by thermal effects in the mirrors to the EC beam with a counteracting effect. By introducing a plate in transmission, placed at Brewster's angle in the collimated arm of the EC, we found that thermal lensing (induced by the temperature-dependent refractive index) in the plate introduces a wavefront distortion with an opposite sign to that caused by the mirrors. By convective cooling with a flushing gas, the thermal lens in the plate can be continuously adjusted. Harnessing this effect, we demonstrated a high-power EC with largely constant beam profile, up to powers exceeding 150 kW, see Fig. 2.3.

Fig. 2.3 **a** Dependence on beam size of the circulating power in the EC. BP: Brewster plate. For a convectively cooled BP, a nearly constant beam size is achieved. **b** Deviation of the beam in the experiment from a perfect Gaussian profile. Adapted with permission from [7] © The Optical Society

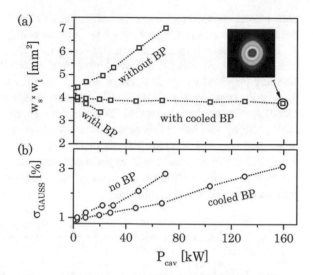

These results represent record powers for ECs including transmissive optics. Possible applications include reflective output coupling of intracavity generated vacuum- or extreme-ultraviolet radiation, and power scaling of free-space soliton ECs (see Sect. 3.3).

2.2 Femtosecond Enhancement Cavities in the Nonlinear Regime [8]

Within the highly nonlinear process of high-order harmonic generation (HHG) in a gas target, the ionization of gas atoms is a crucial step. Compared to the number of electrons recombining and leading to the emission of coherent (extreme) ultraviolet radiation, a much higher number of free electrons are generated in the target via ionization. The number of these electrons strongly increases while the pulse passes through the gas target, which modulates the refractive index of the latter on the same time scale. This leads to a self-phase modulation (SPM), considerably affecting the spectrum and the phase of the pulse as it passes through the gas target.

Including this nonlinear interaction in a EC strongly affects the response of the passive optical system in that the nonlinear distortion of the spectrum and of the phase of the circulating pulse renders the interferometric overlap at the input coupler strongly power-dependent. Commonly, ECs for HHG are optimized for a flat spectral finesse and phase over the spectrum of the seeding pulses, without accounting for this nonlinear distortion—in fact, to date all EC-HHG systems reported in literature use this "standard-approach". For these cavities, the nonlinear interaction of the circulating pulse with the gas target manifests as an intracavity peak power clamping, with increasing seeding power. In particular, this limits the achievable intensity at the HHG focus, restricting both the XUV photon energy and photon flux attainable.

The nonlinear response of standard-approach ECs housing a gas target was observed and qualitatively explained by several groups [9–12]. However, the understanding provided by these works did not suffice for a global optimization of the intracavity conversion process. In [8], we closed this gap by deriving an ab-initio model for the nonlinear response of a standard-approach resonator housing a gas target, and by validating this model over a wide parameter range using a spectral phase measurement previously developed in our group [13]. In Fig. 2.4, examples of the agreement of our model with experimental data are given. Notably, an empirical, simple formula for the "clamping intensity" as a function of the relevant experimental parameters was derived in this work:

$$I_{CL}(\tau, \mathcal{F}, nl) = I_0 \times \left(\frac{\tau_0 - \alpha}{\tau - \alpha} \frac{\mathcal{F}_0 - \beta}{\mathcal{F} - \beta} \frac{n_0 l_0 - \gamma}{nl - \gamma} \right)^{\delta}. \qquad (2.2)$$

Here, I_{CL} is the *clamping intensity*, defined as the peak intensity in the gas target, at which the peak power enhancement equals 65% of its value for linear enhancement (i.e. without the gas target). Furthermore, τ, $\mathcal{F}$, and nl are the (input) pulse duration, cavity finesse and gas-density-length product, respectively. The empirical parameters on the right-hand side of Eq. (2.2) are given in Table 1 of [8]. Importantly, for given gas and pulse parameters and for a target intensity, there is an optimum cavity finesse that minimizes the required input peak power.

This work had two major implications. Firstly, via Eq. (2.2) it provided direct and simple means of optimizing cavity-enhanced HHG for standard-approach ECs, facilitating the choice of the finesse, geometry etc. for a target XUV spectrum.

Secondly, it provided the theoretical apparatus to tackle the question whether the conversion efficiency in EC-HHG could be improved beyond that achievable with standard-approach ECs by accounting for the nonlinearity of the gas target when designing the spectral finesse and phase of the EC. In [8] we discussed the general idea to keep the newly generated (via SPM) spectral components (i) inside of the EC by tailoring the reflectivity of the input coupler and (ii) in phase with the input coupled light by tailoring the spectral phase of the EC mirrors. Simulations predicted that in doing so, a considerable temporal pulse compression should be achievable and, equally importantly, the clamping intensity for standard-approach ECs could be significantly overcome.

A few years later, this idea led to the first demonstration of temporal dissipative solitons in free-space ECs housing a Kerr nonlinearity [14]. This work is described in detail in Sect. 3.3. Here, we should only mention that this new regime for ultrashort laser pulses in passive, nonlinear resonators implemented either in conjunction with an intracavity gas target or by using solely the nonlinearity of the latter, might spawn new regimes of EC-HHG, with unprecedented conversion efficiencies.

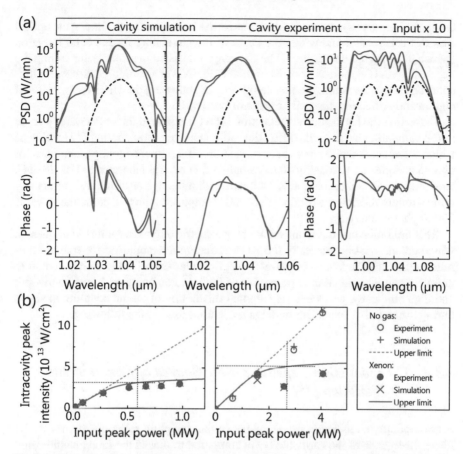

Fig. 2.4 Comparison of experimental data with our ab-initio model for the nonlinear response of standard-approach ECs housing a gas target. **a** Measured power spectral densities (PSD, upper panel) and acquired phases (lower panel) of the intracavity pulse along with simulation results. Left to right: input pulse duration, 640 fs (upchirped), 180 fs (Fourier-limited), 30 fs (fiber-broadened). Without the nonlinear ionization of the gas target (i.e., at low seeding powers), the spectrum is evenly enhanced around the central wavelength. **b** Intracavity peak intensity as a function of input peak power for 640-fs (left) and for 30-fs pulses (right). The outlier in the right panel is due to an incorrectly set offset frequency of the seeding comb. Gas density: 9×10^{18} cm^{-3}; interaction length: 180 μm; finesse: 1190 (narrowband case) and 950 (broadband case). Adapted with permission from [8] © American Physical Society

2.3 Geometric Output Coupling of Intracavity Generated High-Order Harmonics

For almost one decade after the first demonstrations of EC-HHG [15, 16], the extension of this technique to an XUV parameter range suitable for applications in attosecond-resolution photoelectron spectroscopy was hindered by the technological challenge of coupling out high-photon-energy XUV radiation from ECs. In most cases, the harmonics generated at the focus form a beam propagating along the optical axis of the resonator and within the beam of the circulating fundamental radiation (which has a higher divergence). The challenge consisted in spatially separating the high-energy XUV photons from the high-average-power and high-peak-power ultra-short driving pulse before the first cavity mirror following the HHG focus, without significantly affecting the cavity finesse or bandwidth.

Extracting the harmonics via reflection off a plate placed under Brewster's angle in the circulating beam or via diffraction off a nanostructured grating [17] acting as a high-reflectance cavity mirror have been successful output coupling techniques for photon energies up to roughly 40 eV, employed in several laboratories [10, 18–24]. However, the interaction of the XUV beam with reflective or diffractive intracavity optic strongly restricts the bandwidth of XUV output coupling, in particular towards higher photon energies.

This limitation can be circumvented by geometric output coupling techniques, in which geometrical properties of the XUV radiation are employed for spatial separation rather than reflective or diffractive optical elements. Due to their potential to uniquely combine the desired properties of high photon energies, broad bandwidth and high efficiency, we extensively studied this family of output coupling methods during the past five years. Our findings are summarized in the following.

2.3.1 Compact High-Repetition-Rate Source of Coherent 100 eV Radiation [25]

A conceptually straightforward geometric output coupling technique harnesses the lower divergence of the harmonic beam compared to that of the fundamental beam propagating in the TEM_{00} transverse mode of the resonator. In this setting, the harmonic beam can be partially coupled out through a small on-axis aperture in the mirror following the HHG focus [26].

However, drilling a hole with a diameter on the order of 100 μm through a substrate with a thickness of several mm without considerably reducing the optical quality of the substrate surface, proved technologically challenging. Employing inverse laser drilling [27] afforded the first pierced mirrors with a quality sufficient for this application. With these mirrors, we demonstrated geometric output coupling for the first time, boosting the photon energies of MHz-HHG to more than 100 eV, see Fig. 2.5.

2.3.2 High-Harmonic Generation at 250 MHz with Photon Energies Exceeding 100 eV [6]

In a follow-up experiment, building on the results discussed in Sect. 2.1, we implemented advanced cavity designs exhibiting large spots (between 3 and 6 mm

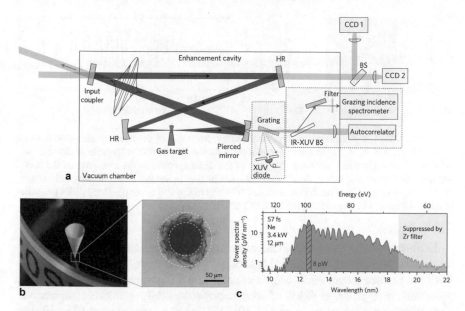

Fig. 2.5 Geometric output coupling of intracavity generated high-order harmonics through an on-axis opening in the mirror following the HHG focus. **a** Cavity layout and beam diagnostics. HR: highly reflective mirror, IR-XUV BS: beam splitter for spatial separation of the output coupled XUV beam from the residual near-infrared fundamental beam leaking out through the hole. CCD: cameras used to measure the profile of the transverse mode excited in the EC. **b** Picture of the pierced mirror. Left panel: back side of the pierced mirror, revealing a cone-shaped extension of the round opening, introduced to minimize diffraction losses of the XUV beam. Right panel: front side of the pierced mirror. The edge shows considerable chipping in a range comparable to the radius of the opening. **c** Output coupled harmonics generated in a neon gas target. The photon energies exceeded 100 eV, at that time improving the state of the art in MHz-HHG by about a factor of four—albeit with a rather low average power of less than 8 pW per harmonic. Adapted with permission from [25] © Springer Nature

$1/e^2$-intensity radius) on all mirrors, with carefully chosen mirror substrates and broadband coatings. In an empty cavity without pierced mirror, the implementation of these measures together with employing a state-of-the-art Yb-based femtosecond frontend, led to a 250-MHz-repetition-rate train of 30-fs pulses with an average power of 20 kW, which constituted record intracavity pulse parameters in terms of the power level and duration.

Including a pierced-mirror XUV output coupler, resulted in a decrease of the circulating power to 12.5 kW, mainly due to thermal stress in this optic, as at that time pierced mirrors were only available on fused-silica substrates. Previously observed occasional damage of the coating of the pierced mirror around the aperture, which we attributed to local field enhancement at the chipped edge of the hole (cf. Fig. 2.5b), was mitigated in this experiment by considerably improving the production of pierced mirrors (see Fig. 2.6a).

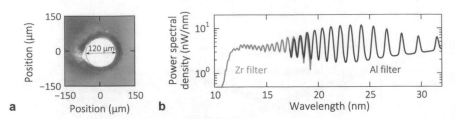

Fig. 2.6 **a** Detail of pierced mirror surface. Notably, the edges of the aperture are smoother than those in Fig. 2.5b. Also, due to the larger spot sizes on the EC mirrors, a larger hole diameter was afforded. This additionally improves the ratio of the clear aperture to the total area on the mirror surface which does not reflect the fundamental beam (i.e. including non-reflecting hole edges), introducing losses to the latter. **b** Power spectral density of the output coupled harmonics generated in neon. Adapted with permission from [6] © The Optical Society

Furthermore, the quantitative understanding of gas-nonlinearity-related intensity limitations [8] allowed for a global optimization of the cavity parameters for HHG. With the pierced-mirror XUV output coupler and a high-pressure neon gas target, 30-fs intracavity pulses at 10 kW of average power in steady operation were reached. This afforded a peak intensity of 3×10^{14} W/cm^2 at a focus size of 10×18 μm^2 and led to the generation of harmonics with photon energies exceeding 100 eV at an unprecedented repetition rate of 250 MHz and with an average power higher by roughly two orders of magnitude than in the previously mentioned work (Fig. 2.6b).

There were three important implications of this experiment. Firstly, for the first time for a multi-MHz source the number of high-energy photons per second was large enough to promise a regime of ~1 photoelectron released from a crystal per pulse in a—space-charge-free—photoelectron spectroscopy experiment.

Secondly, HHG in xenon and argon with the same EC led to XUV power levels comparable to those in the previously discussed work at 78 MHz. Furthermore, acceleration of the atoms of these gases by mixing a lighter carrier gas (helium) led to a reproducible improvement of the XUV power by 30%. These findings prompted systematic studies of both the tradeoff between the efficiency of the HHG process and of output coupling, and of the effect of cumulative plasma in the HHG focus, discussed in Sect. 2.4.

Finally, the repetition frequency in this experiment is still the highest one for HHG to date. This affords outstanding conditions for (direct) frequency-comb spectroscopy and, in particular, in combination with the high photon energies might pave the way to precision-spectroscopy of nuclear transitions, serving, e.g., as a reference for future nuclear clocks [28].

2.3.3   Cavity-Enhanced High-Harmonic Generation with Spatially Tailored Driving Fields [29]

Prior to the work presented in this paragraph, all EC-HHG systems reported in literature employed a fundamental Gaussian mode propagating in the resonator. However, propagation in the optical resonator can also be achieved in other transverse (eigen-) modes while preserving the optical stability of the resonator (i.e., with low diffraction losses) and, thus, without reducing the finesse. Tailoring the transverse mode, in particular in conjunction with geometric XUV output coupling, offers solutions for a number of applications.

In this first work demonstrating EC-HHG with a non-Gaussian beam, the initial motivation was to avoid the high intensity of the circulating beam at the edges of the hole in the XUV output coupling mirror (cf. Fig. 2.5), improving power scalability. Our method, dubbed "quasi-imaging", employs a degeneracy of certain higher-order modes at the center of the resonator stability range to tailor a transverse mode with an on-axis intensity minimum at the pierced output-coupling mirror and an on-axis intensity maximum close to a tight cavity focus, for efficient HHG[1] [30, 31]. In contrast to other solutions proposed in literature [26, 32–35], quasi-imaging is readily implemented in any standard EC with a small on-axis opening in the mirror after the focus, rendering it conceptually simple and robust.

In our proof-of-principle experiment, we realized a coherent superposition of the Gauss-Hermite transverse modes of order $(0, 0)$ and $(0, 4)$ of a bow-tie ring resonator, by (i) operating the resonator at the center of the stability edge, (ii) including an on-axis slit in the mirror following the focus and (iii) suppressing transverse modes of even higher orders with adequate beam blocks. The resulting intensity distribution is shown in Fig. 2.7. With a xenon gas target, we demonstrated a conversion efficiency comparable to that previously achieved with a Gaussian mode and argued that due to the considerably larger opening in the XUV output-coupling mirror, a significantly higher output-coupling efficiency (close to 100%) can be expected, even for lower-order harmonics. Furthermore, we showed that HHG in the intensity maximum after the focus results in a well-collimated, round far-field beam profile.

Besides the demonstrated result that the output coupling efficiency can be increased even for lower-order harmonics, this work had a number of implications, both for enhancement-cavity experiments in general and for our subsequent EC-HHG research in particular. The main property leading to the applications in the first category is the fact that quasi-imaging geometrically provides on-axis access to an optical resonator which can potentially combine a very high finesse (in the absence of significant losses at the aperture) with an on-axis maximum of very high intensity close to the focus. This could for instance be useful for high-harmonic generation spectroscopy [36]: for a thin molecular gas, the SPM-induced intensity clamping

---

[1] The name originates in the property of such a resonator to resonantly enhance more than one transverse eigenmode for a given Gouy phase. However, because not *all* eigenmodes are simultaneously resonant, the optical system is not an imaging one (which would result in an optically unstable system). The name was given by J. Weitenberg.

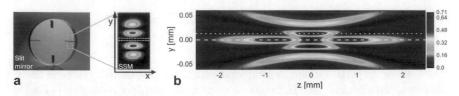

Fig. 2.7 **a** The left panel shows a mirror with four slits, one of which we used to couple out intracavity-generated harmonics. The right panel shows the intensity distribution in the plane perpendicular to the optical axis (x, y) as measured at the surface of the output coupling mirror. The wide area of low intensity is clearly visible. SSM: "simple slit mode". **b** Calculated intensity distribution around the focal region in the (y, z)-plane, i.e. parallel to the optical axis. In the focal plane, the SSM exhibits the same shape as on the output-coupling mirror. However, around one Rayleigh range in front and behind the focal plane, it forms strong on-axis intensity maxima. Adapted with permission from [29] © American Physical Society

limitation is alleviated, affording high peak intensities and, therefore, the disentanglement of the single-/few-molecule dipole response from the collective response of a multitude of molecules (including macroscopic phase matching). The signal is expected to be very weak such that the high repetition rate inherent to ECs would provide ideal conditions for this analysis. Furthermore, the broad bandwidth of this XUV output coupling technique would allow for broadband HHG spectroscopy.

Another application profiting from the particular geometry of a quasi-imaging EC is Thomson (or inverse Compton) scattering of near-infrared photons off relativistic electrons [37]. The scattering of picosecond pulses circulating inside of the EC off counter-propagating electron bunches results in a beam of hard X-rays, which needs to be coupled of the EC. The large opening in the mirror following the high-intensity focus—affordable without reducing the finesse of the EC—might provide an excellent solution to this problem.

For our research, this work had the important implication that it prompted the development of a powerful numerical model for HHG. Besides supporting the findings of this work, this model was later used in various studies, including an ample study of methods suitable for EC-based generation of isolated attosecond pulses [38, 39] (see also Sect. 3.2) and the quantitative investigation of the output-coupling efficiency in EC-HHG with geometric output coupling [40].

2.3.4 Cavity-Enhanced Noncollinear High-Harmonic Generation [39]

In the context of our studies regarding the possibility of generating isolated attosecond pulses via gating mechanisms inside of ECs, we investigated the feasibility of manipulating a TEM_{01} cavity mode. This work is described in detail in Sect. 3.2. Briefly, the main idea is to delay one lobe of this mode with respect to the other one by an odd number of half-cycles before the mirror focusing the beam onto the HHG target. In this manner, an on-axis intensity maximum is achieved, optionally with a wavefront rotation in the focal region. This phase shift between the two lobes can be reversed after the focus, enabling propagation with low diffraction losses. We refer to this mode of operation as *transverse mode gating* (TMG).

The ability to generate isolated attosecond pulses via TMG is discussed in Sect. 3.2. Here, we restrict the discussion to the case where the phase shift between the two lobes is chosen to be π at the central wavelength. With TMG, an intensity maximum is achieved in and around the focal plane, resulting in the emission of high harmonics along the optical axis of the resonator (see also Sect. 3.2). Like the quasi-imaging mode, the TMG mode allows for a wide opening in the mirror after the focus, enabling high output coupling efficiencies. However, in TMG the position within the stability range can be freely chosen, affording power-scalable designs with large spot sizes on the EC mirrors (see Sect. 2.1).

A comparison between the output coupling efficiency with the standard pierced-mirror approach for the TEM_{00} mode and TMG is shown in Fig. 2.8. Importantly, in contrast to the TEM_{00} mode with output coupling through an on-axis hole, for gas target positions almost over the entire range of a Rayleigh length before and after the focus, the output coupling efficiency with TMG exceeds 50%, affording attractive prospects for increasing the overall conversion efficiency (XUV output coupled power divided by NIR input power) in EC-HHG.

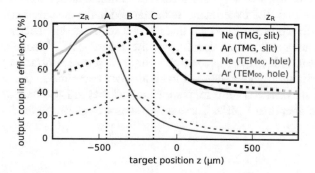

Fig. 2.8 Bold, black lines: calculated output coupling efficiency with the TMG mode for HHG in argon (harmonic 33 at a peak intensity of 1.5×10^{14} W/cm², dashed lines) and neon (harmonic 79 at a peak intensity of 3.0×10^{14} W/cm², solid lines), with 40-fs pulses centered at 1025 nm, as functions of the gas target position. For comparison, the calculated output coupling efficiency with the TEM_{00} mode and a pierced-mirror output coupler is shown (thin red lines), assuming the same HHG parameters and dimensions of the opening in the output coupling mirror leading to round-trip losses of 1% in both cases. The grey continuations of the curves identify when pulse energies >80 µJ are needed to reach the corresponding peak intensities in our experiment and the vertical dotted lines (A, B and C) mark particular gas target positions in our experiment (see manuscript for details). Adapted with permission from [39] © The Optical Society

2.4   The MEGAS Beamline

The project MEGAS, *Megahertz Attosecond Pulses for Photoemission Electron Spectroscopy and Microscopy* (see Sect. 1.2) provided an ideal framework for the development of a new attosecond beamline combining all of our insights on cavity-enhanced HHG, and optimized for long-term stable operation—the MEGAS beamline. The beamline layout is sketched in Fig. 2.9 and depicted on the subsequent pages. This section reviews three major milestones on the road to the first high-photon energy, high-flux and multi-MHz-repetition-rate attosecond-resolution photoelectron spectroscopy (PES) experiments.

These milestones were (i) setting up a 100-W-class source of near-infrared ultrashort pulses with a high control over the pulse waveform and with the possibility of tuning the pulse repetition frequency. The former requirement ensured optimum conditions both for stably seeding an EC over several hours of continuous measurement time and provided necessary prerequisites for the generation of temporally isolated attosecond pulses (IAP).

The latter requirement enabled (ii) the first direct, experimental study of cumulative plasma effects in HHG in gases, revealing a dramatic efficiency drop for plateau harmonics in the multi-shot case and, therefore, influencing the final design of the beam line. In addition, adapting the repetition frequency of the system allowed us to optimize time-of-flight-based photoelectron detection to nearly 100% temporal duty cycle in attosecond-PES experiments (summarized in the next section).

Finally, (iii) we investigated the tradeoff between the efficiencies of the HHG process and geometric output coupling of the coherent XUV radiation, in theory and experiment. This allowed us to reach—for the first time—intracavity conversion efficiencies similar to those achieved in single-pass experiments (i.e. without EC), and the highest usable XUV flux for photon energies beyond 40 eV at pulse repetition frequencies higher than 1 MHz. These milestones are detailed in the following.

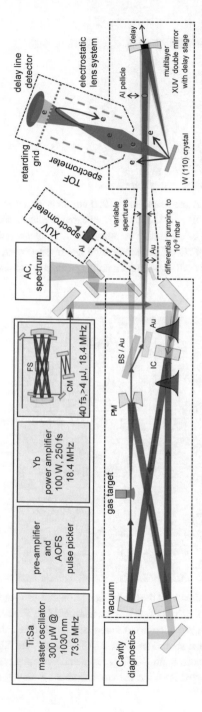

Fig. 2.9 The initial pulse train of the beamline is generated by a Ti:Sa oscillator with a fundamental repetition frequency of 73.6 MHz and with an output spectrum optimized to reach a power level in a band around 1030 nm sufficient for seeding a chain of Yb-doped fiber amplifiers. The fiber-based chirped-pulse amplification (CPA) system delivers an output train of 250-fs pulses with a power of more than 100 W at a repetition rate tunable down to 18.4 MHz by means of an acousto-optic-frequency-shifter (AOFS) based pulse picker at the beginning of the amplifier chain. A newly developed temporal pulse compression unit based on spectral broadening via self-phase modulation upon multiple passes through bulk fused silica (FS) and dispersion-compensating chirped mirrors (CM) provides the seeding pulses for the enhancement cavity. The choice of the input coupler (IC) and geometry (including pierced mirror, PM) for the EC are optimized for highest overall conversion efficiency of HHG in argon. Diagnostics of the intracavity circulating pulses include a spectral measurement, autocorrelator (AC), transverse beam profiler and power meters. The EC is operated close to the inner stability edge with a mode size of 4.2×3.1 mm^2 ($1/e^2$-intensity diameter) on the mirrors, resulting in a focus size of 25×3.1 μm^2. The two parameters of the seeding frequency comb were adjusted to ensure a long-term steady coupling to the EC, see Saule et al., Methods[161]. The harmonics coupled out through the 340-μm opening in the spherical mirror after the focus can be characterized with XUV diagnostics (monochromator, calibrated photodiodes), or guided through a differential vacuum pumping system to the high-vacuum experimental chamber for PES. For pump-probe experiments, the portion of co-propagating near-infrared light leaving the EC towards the PES chamber can be tuned by the choice of the beam splitter (BS)/gold (Au) mirror following the PM. The PES experimental chamber and benchmarking measurements are detailed in the next section. Adapted with permission from [45] © Springer Nature

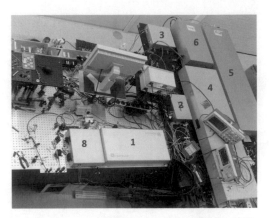

Femtosecond frontend. 1: Ti:Sa master oscillator, 2: pre-amplifier at 1030 nm, fiber stretcher, 3: feed-forward carrier-envelope phase control for 1030 nm, 4: optional fiber-based nonlinear pulse compression, 5, 6: Yb-fiber power amplifier (2 stages), 2 grating compressors, 7: nonlinear pulse compression to <40 fs at an output pulse energy of >4 µJ and a repetition rate of 18.4 MHz, 8: feed-forward carrier-envelope phase control for the 78-MHz, 6-fs pulses exiting the master-oscillator

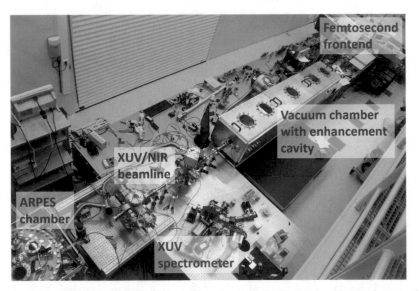

Top view of the MEGAS experimental setup. The pulses from the femtosecond frontend are coupled to the 8-mirror femtosecond enhancement cavity located in the 3-m-long vacuum chamber. The XUV and a copy of the NIR pulse driving HHG co-propagate through a differentially pumped beamline to the high-vacuum ARPES chamber containing the sample under test. Alternatively, the XUV beam can be separated from the NIR beam and guided to an XUV spectrometer. On this picture, the beamline and XUV spectrometer are disconnected from the vacuum chamber containing the enhancement cavity

Inside view of vacuum chamber with 4 enhancement cavity mirrors. In front of the mirrors, metal masks are mounted, to prevent stray light from impinging on the mirror mounts, which can misalign the resonator due to thermal deformations

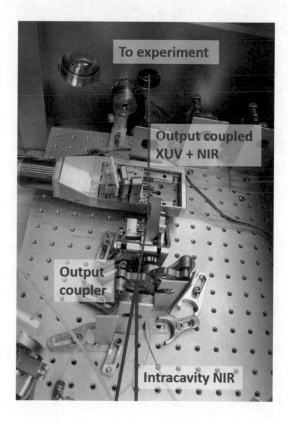

Temporally synchronized XUV (dashed violet line) and NIR (red line) radiation is transmitted through the pierced output coupling mirror and guided via an unprotected gold mirror placed under grazing-incidence, to the to the (AR) PES experiment. Optionally, a dielectric-coated glass plate can be driven in the output coupled beam, with a high transmission for the NIR and a high reflection for the XUV radiation. The power of the latter can then be measured after an additional XUV transmission filter with a calibrated photodiode

Close-up view of the end-fire gas nozzle mounted onto an xyz kinematic stage, located inside the vacuum chamber containing the enhancement cavity

Close-up view of the gas nozzle in operation. The cavity focus is very close to the opening of the nozzle, where the velocity of the exiting atoms is highest. While the ionized argon atoms propagate away from the nozzle opening, they emit fluorescence, visible as a bright, "white cloud". Near the focus, the cavity beam is intense enough to ionize atoms from the background gas in the chamber (in this case the pressure was ~10^{-2} mbar). Further away from the gas target, two sections of the cavity beam become visible. We attribute this to the interaction of the high-power beam with excited-state atoms

2.4.1 *Phase-Stable, Multi-µJ Femtosecond Pulses from a Repetition-Rate Tunable Ti:Sa-Oscillator-Seeded Yb-Fiber Amplifier [41]*

At the time of planning the beamline, modelocked titanium-sapphire oscillators with subsequent acousto-optic-frequency-shifter (AOFS) based feed-forward carrier-envelope-phase (CEP) stabilization provided an optimum combination of robustness and high CEP stability [42]. In addition, this frontend technology was commercially available, and the direct output of these systems readily delivered waveform-stable, few-cycle pulses, suitable as an ultrashort pump of attosecond dynamics in solid targets (e.g., for nanoplasmonics).

Our Ti:Sa oscillator was optimized to produce 300 µW of average power in a 10-nm band around 1030 nm, sufficient to seed the subsequent Yb-based amplifier

Fig. 2.10 Power spectral density (PSD) of the phase noise and integrated phase noise (IPN) in the frequency band 0.4 Hz–400 kHz, for four different repetition rates. Adapted with permission from [41] © Springer Nature

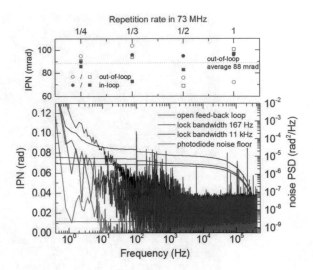

system. An AOFS feed-forward CEP stabilization unit was designed for the central wavelength of 1030 nm, and a novel pulse picker technique [43], preserving the CEP stability of the input pulse train, was developed in collaboration with the MEGAS project partners. The Yb-fiber power amplifier delivered linear amplification for 250-fs pulses with more than 80 W of average output power at the fundamental repetition rate of 74 MHz and at 1/2, 1/3, and 1/4 of the latter (with a corresponding increase in pulse energy). Using standard fiber-based nonlinear compression at a pulse energy of 0.6 μJ, and feeding the CEP stabilization with an error signal combining contributions of CEP fluctuations from the oscillator and from the power amplifier, we measured an (out-of-loop) phase noise of the entire system below 100 mrad, integrated in the frequency band between 0.4 Hz and 400 kHz for all repetition frequencies, see Fig. 2.10.

For the final configuration of the MEGAS beamline (see Fig. 2.9), operated at 18.4 MHz, a customized pulse compression unit was developed [44], delivering sub-40-fs pulses with an energy of 4.5 μJ and largely preserving the high phase stability [45].

2.4.2 Cumulative Plasma Effects in Cavity-Enhanced High-Order Harmonic Generation in Gases [46]

Despite providing ultrashort circulating pulses with multi-kW-level average powers, the overall conversion efficiencies demonstrated with EC-HHG—i.e. output XUV power divided by input power—have so far remained beneath the record values, in the order of 10^{-7}–10^{-6} achieved with single-pass systems with comparable gas targets at lower repetition rates (see, e.g. overview in Ref. [40]). Three EC-related limitations can be made responsible for this. Firstly, in standard-approach ECs, the affordable ionization fraction per pulse is strongly restricted by the self-phase-modulation-induced intensity clamping, and rapidly decreases with increasing finesse (see Sect. 2.2).

Secondly, at pulse repetition periods on the order of 10 ns, the gas target is normally not entirely replenished between subsequent pulses, leaving a substantial fraction of the atoms ionized or in excited states. This potentially affects both the longitudinal phase matching conditions and the transverse spatial conditions for harmonic emission. The MEGAS frontend provided ideal prerequisites for a direct study of these cumulative plasma effects, as summarized in this section. The third limitation of the overall conversion efficiency in broadband EC-HHG regarded the efficiency of coupling out harmonics through a pierced mirror, and is discussed in the next section.

Setting up an 18-MHz fundamental-repetition-rate EC with a tight focusing, and making use of the pulse picker in the frontend (see Fig. 2.9), allowed us to compare EC-HHG in argon at 18 and 36 MHz with otherwise identical parameters of the individual seeding pulses (0.6 µJ energy and 35 fs duration). In this experiment, we systematically varied the gas target position as well as the backing pressure for two regimes: in the single-pulse regime (SP), at 18 MHz, in good approximation each gas atom was hit by one pulse only; in the double-pulse regime, at twice the repetition rate (and, thus, average power), each atom was hit twice. The results are shown in Fig. 2.11.

Intriguingly, for most harmonics the overall flux is very similar for both repetition rates, indicating a dramatic drop in efficiency in the regime where multiple pulses hit the gas target, see Fig. 2.11. For the two repetition rates, the optimum generation conditions (target gas density and position) are very similar, see Fig. 2.11a. This indicates that the ionization fraction in the part of the target gas that contributed to HHG is similar for both the SP and the DP (because a different ionization fraction would significantly change the phase-matching pressure), and the substantial decrease in efficiency can most probably be attributed to a decrease in the generation volume: in the DP regime, the gas target is spatially partitioned into a part that was hit by the previous pulse, contributing little to the flux (e.g., due to pre-ionization or excitation), and a part with "fresh" ground-state gas atoms. This would affect the output coupled flux in two ways: firstly, the number of atoms contributing to HHG is decreased, and secondly, the reduced generation volume results in a larger divergence of the harmonic beam and, thus, in a lower output coupling efficiency. To confirm/disentangle these effects, a similar experiment with a Brewster plate or a diffraction-based output coupling method (see references in the introduction to Sect. 2.3) could be performed in future.

The overall conclusion of these investigations is that for gas-HHG a regime, in which each atom interacts with a single pulse is highly desirable. Very similar observations were made in a concomitant, independent experiment [23]. The latter work reported technological measures to accelerate heavier gas atoms both by mixing them with lighter atoms and by strongly heating the gas nozzle, demonstrating mW-level powers for harmonic around order 13 and repetition rates of ~80 MHz.

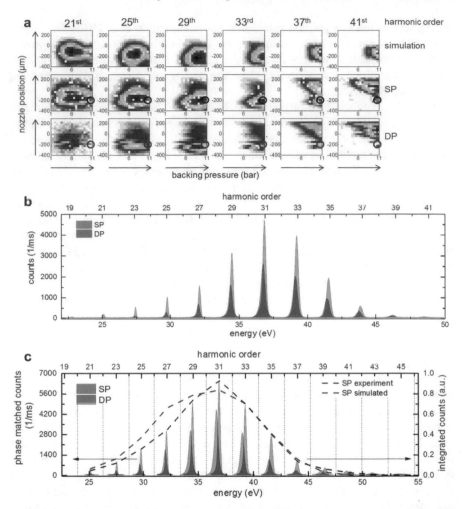

Fig. 2.11 **a** XUV flux measured as a function of the gas target position and backing pressure, for six different harmonics, in single-pulse (SP, 1st row) and double-pulse (DP, 2nd row) regime. For each harmonic, the color scales are common to the SP and DP maps; they are normalized to the overall highest flux. The nozzle position of 0 μm marks the cavity focus. **b** XUV spectra for both repetition rates measured for the parameters indicated in **a** by the circles. **c** Stitched XUV spectrum for individually optimized conditions for SP and DP, taken from the positions on the maps with the highest flux. The dots represent the integrated flux within these harmonics. Adapted with permission from [46]

2.4.3 Efficiency of Cavity-Enhanced High-Harmonic Generation with Geometric Output Coupling [40]

To support the theoretical understanding of cavity-enhanced HHG, a comprehensive numerical model [38] for HHG was developed[2] in our group during the last few years. This model numerically solves the first-order propagation equation for the IR and XUV fields, with optional envelope approximations, and uses the strong-field approximation [47]. Importantly, the model can be applied to cases where rotational symmetry is not given. It accounts for plasma-induced lensing, absorption and spectral blue-shift, as well as for Kerr focusing and self-phase modulation.

While we found the cumulative plasma effects (see previous section) too complex for a quantitative simulation, very good agreement between our model and experimental results obtained in the single-pulse regime was obtained over a large range of parameters, as discussed in the paper summarized in this paragraph.

Using this model, a simple analytical formalism for the divergence of the intracavity harmonic beam as a function of experimental design parameters such as gas target position, cavity geometry and driving pulse intensity, was derived in this work, thereby establishing a connection between the measured XUV spectra and the macroscopic response of the intracavity nonlinear medium. A trade-off between the efficiency of geometric output coupling and that of the HHG process is elucidated and the significant share of the output coupling efficiency to the overall HHG conversion efficiency is quantitatively provided. Together with the previously developed understanding of plasma-related enhancement limitations (see Sect. 2.2), this model provided a holistic means of optimizing the overall efficiency of EC-HHG employing geometric output coupling, which played a crucial role in the final design of the MEGAS beamline.

2.5 High-Flux Ultrafast Extreme-Ultraviolet Photoemission Spectroscopy at 18.4 MHz Pulse Repetition Rate [45]

2.5.1 HHG Source

The design criteria discussed in the previous paragraphs allowed the efficient use of the state-of-the-art, 100-W femtosecond frontend [41] for cavity-enhanced HHG. The repetition rate of the system was set to 18.4 MHz, as a good compromise among the following criteria:

- For photoelectron spectroscopy, we used a time-of-flight detector (ToF) with a drift distance of 880 mm (Themis 1000, SPECS GmbH). At our repetition period

---

[2] The numerical model was primarily developed by M. Högner, with contributions from V. Yakovlev, V. Tosa and I. Pupeza.

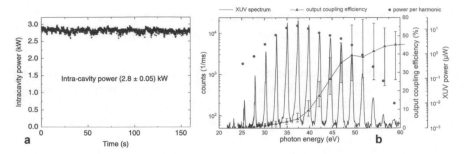

Fig. 2.12 **a** Measurement of the intracavity power level over a duration of 160 s, under experimental conditions for optimized HHG. The root-mean-square of the intensity fluctuations amounts to 1.8%. **b** Output coupled XUV spectrum, power per harmonic, and output coupling efficiency. Adapted with permission from [45] © Springer Nature

of 54 ns, this results in a detectable window of 33–50 eV electron kinetic energy with a resolution of around 30 meV.

- The frontend pulse energy of roughly 6 μJ enabled the efficient use of a multi-pass nonlinear compression stage to sub-40-fs pulses which is a pulse duration readily supported by the bandwidth of state-of-the-art cavity mirrors at 1-μm wavelength (achieving significantly shorter pulse durations in ECs currently is technologically challenging, as discussed in the next chapter). The input pulses to the EC had an energy of more than 4 μJ leading to intracavity circulating pulses with more than 150 μJ for HHG, at a moderate cavity finesse.

- Along with the reduced cumulative effects due to the relatively low repetition rate, the latter resulted in an unprecedented efficiency in the photon energy range between 20 and 60 eV.

Very important for experiments extending over several minutes is the absence of drifts (and lock interruptions) over these time spans. Figure 2.12a shows a typical measurement of the intracavity power level over a few minutes. The system was regularly operated for more than 30 min without the need of re-locking or realignment. Hour-scale drifts due to the high average power were unproblematic for the proof-of-principle PES experiments performed and can be removed by improved high-power beam handling, if necessary in the future.

Thanks to the knowledge of the output coupling efficiency (see previous section and Fig. 2.12b), we inferred a record intracavity HHG conversion efficiency, with a maximum in the order of 10^{-7} for photon energies between 35 and 40 eV, approaching that of single-pass HHG in argon. This provided a unique combination of high repetition rate (18.4 MHz), high photon energies (up to 60 eV) and high usable XUV powers (up to 20 μW in the strongest harmonics between 35 and 40 eV), see Fig. 2.12b. In the landscape of contemporary HHG sources, this combination provided unique prerequisites for space-charge-free, high-flux, attosecond time-resolution photoemission electron spectroscopy (PES), see Fig. 2.1a.

2.5.2 Laser-Assisted Photoemission Electron Spectroscopy at 18.4 MHz—Photoelectron Statistics

To demonstrate the benefits of our high-repetition-rate HHG source for space-charge-free PES, we guided the XUV beam (and, optionally, the co-propagating NIR beam) through a differentially-pumped beamline, and focused the radiation with a standard delay mirror onto a 10-μm-diameter spot size on the surface of a tungsten crystal, see Fig. 2.9.

Static PES measurements (i.e. with exclusively XUV illumination) yielded a number of 10^{10} photoelectrons per second emitted from the sample surface, at our repetition rate corresponding to estimated space-charge-induced distortions in the photoelectron spectra of merely 10 meV, see Fig. 2.1b. Importantly, the XUV photons generating photoelectrons at this rate had energies of up to 60 eV. These photon energies are high enough to liberate electrons from more tightly bound core states in addition to valence-band electrons, allowing for self-referencing delays in photoemission from two different states in the same measurement. For photon energies higher than 40 eV, at identical space-charge-induced distortions, the count rate in our experiments exceeded those reported in literature for state-of-the-art attosecond-PES by up to 3 orders of magnitude, see [45].

We investigated the stability of our system by acquiring a large number of photoelectron spectra, see Fig. 2.13a. We found that the system behaves statistically (standard deviation decreases with the square root of measurements) up to a number of 2.9×10^9 shots, reaching a standard deviations of the photoelectron spectra of less than 1%. At our repetition rate, this corresponds to a measurement time of 160 s. For comparison, at a repetition rate of 10 kHz, typical for state-of-the art attosecond-PES experiments, this would correspond to 3.4 days. Furthermore, a standard deviation of ~2.5% was inferred for 1.7×10^{10} laser shots, corresponding to 15 min at 18.4 MHz and to 19 days of continuous measurement at 10 kHz.

In the presence of a copy of the NIR field generating the harmonics, phase-locked to them and variably delayed, sidebands appear in the photoelectron spectrum at energies between those of the photoelectrons emitted by the individual harmonics alone [48]. In this "laser-dressed" PES setting, the sidebands are modulated as a function of the delay between the NIR pulse and the XUV pulse train, and information on the participating light fields and on electron dynamics in the sample (such as, e.g., photoionization delays from different states) can be gained from the interpretation of the relative phase of these sideband modulations [49–55]. Figure 2.13b shows such a sideband modulation obtained with our system, demonstrating the capability of attosecond-resolved PES, further elucidated in the next section.

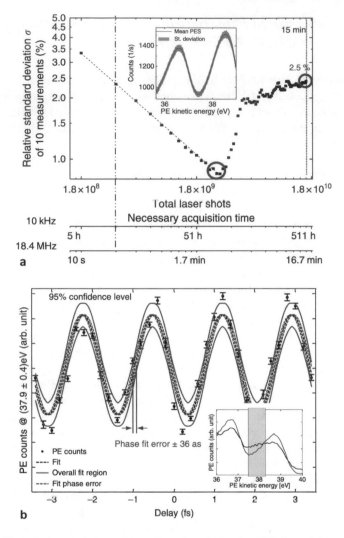

Fig. 2.13 Photoelectron statistics and time-dependent sideband modulation. **a** Measurement over
~15 min (1.8 × 10^10 laser shots). The plot shows the evolution of the relative standard deviation σ
of the counts (averaged from 35 to 44 eV kinetic energy). The dashed line is a $1/\sqrt{T}$-fit to the first
160 s. The excellent agreement confirms statistical behavior up to 160 s with a minimum relative
standard deviation of 0.9% for this measurement time (red circle). For $T > 160$ s, slow drifts emerge.
The inset shows the mean (black) and relative standard deviation (red) of PES measurements with
160 s integration time. The bottom axes illustrate the time a 10-kHz system would need for the
same amount of laser shots (under identical space-charge limitations). **b** Intensity of the sideband
at 37.9 eV kinetic energy as a function of the NIR-XUV delay taken within 105 s in total. The
phase fit error (red dashed line) corresponds to 36 as and is comparable to the timing jitter of the
interferometer (see original paper). Blue lines show the maximum and minimum error boundaries
of all fit parameters. The inset shows photoelectron spectra at two different delays with (black) and
without (red) sideband and the spectral region over which the sideband intensity was integrated.
The error bars are calculated by dividing the dataset into 10 subsets, calculating the relative standard
deviation of the 10 resulting data points at each delay step and dividing it by $\sqrt{10}$ to account for
the better statistics of the entire dataset. Adapted with permission from [45] © Springer Nature

2.5.3 Attosecond Angle-Resolved Photoemission Electron Spectroscopy (Attosecond-ARPES) at 18.4 MHz

At the time of writing this book, the MEGAS beamline was being used by the group of Prof. Ulf Kleineberg for attosecond-ARPES measurements at 18.4 MHz [56].

Figure 2.14 shows typical measurements of a (110)-tungsten crystal, albeit after integration over the angle dependence of the data. Thus, the kinetic energy of the photoelectrons versus the NIR-XUV delay is shown.

For the PES measurements shown in Fig. 2.14a, an argon gas target was used for cavity-enhanced HHG (cf. XUV spectrum in Fig. 2.12b). The energy of the XUV photons is large enough to overcome the work function ϕ_{W110} of the tungsten-110 crystal (5.25 eV) and accelerate photoelectrons released from the valence band to the detector. The kinetic energy E_{kin} with which a photoelectron released by a photon with the energy E_{phot} leaves the sample is [57]:

$$E_{kin} = E_{phot} - \phi_{W110} - |E_B|, \tag{2.3}$$

where $|E_B|$ denotes the binding energy of the electron inside the crystal. In addition, in an ARPES measurement, the momentum with which the electron leaves the crystal can be inferred from the measured emission angle (not shown here).

While conceptually attosecond-temporal-resolution PES measurements require a broad XUV spectrum according to Heisenberg's uncertainty principle, two-photon, two-color PES employing *attosecond pulse trains* (APT) offers an elegant route towards combining a high spectral resolution with a high temporal resolution, with certain constraints [55]. On the one hand, the spectra of APT, generated by relatively long (i.e., multi-cycle) VIS/NIR pulses, exhibit individual harmonics, arising from the interference of the XUV emission from each contributing half-cycle (cf. XUV spectrum in Fig. 2.12b). The narrower the individual harmonics, the higher the spectral resolution of the PES measurement.[3]

On the other hand, the temporal coherence of the HHG process preserves the timing of the APT with the driving VIS/NIR electric field, which underlies the ability to measure relative photoionization delays with sub-optical-cycle timing accuracy.[4] At the sample, in the presence of a copy of the field generating the APT via HHG (in our case of the photon energy $E_1 = 1.2$ eV corresponding to a central wavelength of 1030 nm), photons from two adjacent harmonics with the photon energies

---

[3] Note the analogy to frequency combs: the frequency-domain structure of the train of pulses emitted by a (phase-stabilized) modelocked oscillator is that of a set of equidistant laser lines (see, e.g., Sect. 1.2).

[4] The agreement of these measurements with Heisenberg's uncertainty principle is noteworthy: while photoionization delays can be measured with attosecond accuracy, in such a measurement it is impossible to assign the half-cycle of the VIS/NIR field, at which a certain photoionization event occurred. Therefore, the temporal uncertainty extends over the duration of the optical pulses involved.

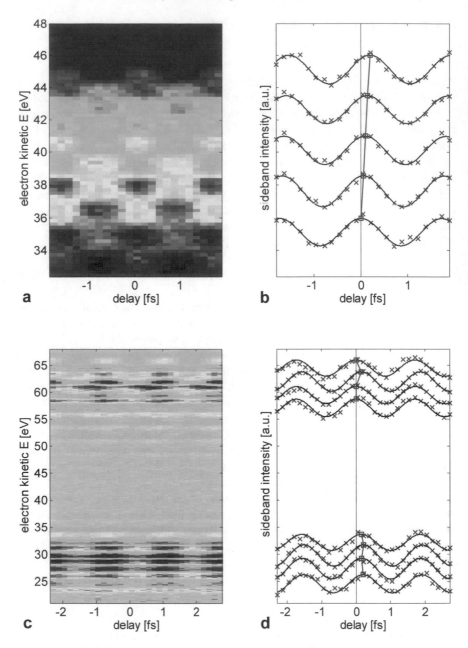

◄**Fig. 2.14** Attosecond-ARPES measurements (courtesy of Prof. Ulf Kleineberg). **a** Angle-integrated photoelectron spectra of a (110)-tungsten crystal, recorded as a function of the delay between the 1030-nm-centered laser and the XUV pulse train generated via HHG driven in an argon gas target; 23 delay steps of 50 nm were recorded within 57 s. The sideband modulations contain information on the timing of photoemission (see text). **b** Lineouts through the sidebands in the spectrogram in (**a**), corresponding to photon energies of even harmonic orders 34, 36, 38, 40 and 42. The blue line follows the central maximum of cosine functions fitted to the sideband oscillations, indicating their relative phase. **c** Measurements of the same sample, performed with higher photon energies, obtained by using a neon gas target for HHG. Due to the lower XUV power generated with neon, for 31 delay steps of 50 nm each, a total measurement time of 3 h was necessary to obtain this spectrogram. **d** Lineouts through the sidebands (here: 52, 54, 56 and 58) obtained both from valence-band electrons and from the 4f core state

E_{2q-1} and E_{2q+1}, each together with a photon of energy E_1 can contribute to two-photon photoionization. The two quantum paths described by the total photon energy $E_{2q-1} + E_1$ (absorption of a photon with E_1) and $E_{2q+1} - E_1$ (stimulated emission [58] of a photon with E_1) lead to the same final energy state and the photoelectron wavepackets interfere according to the phase difference of the two quantum paths, which depends on the delay t between the fundamental field and the APT [48]. In the photoelectron spectrum this two-photon, two-color photoionization leads to "sidebands" centered at kinetic energies corresponding to even multiples of the fundamental photon energy, with an intensity depending on the relative phase of the interfering electron wavepackets. The t-dependent modulation of the intensity I_{2q} of the sideband of order $2q$ is in good approximation given by the relation:

$$I_{2q}(t) \propto \cos(2\omega_{IR}t - \Delta\varphi_{XUV} - \Lambda\varphi_{at}), \tag{2.4}$$

where ω_{IR} is the angular frequency of the fundamental (near-infrared) field, and $\Delta\varphi_{XUV}$ and $\Delta\varphi_{at}$ are two q-dependent phase values that reflect photoemission delays due to relative phases of the individual harmonics (at the sample) and photon-energy-dependent photoemission delays from the material system, respectively. Note that the appearance of strong sidebands of the two-photon photoionization depletes the strength of the single-XUV-photon photoionization bands (Fig. 2.14a). A fit of such cosine functions to the sideband modulations shown in Fig. 2.14a is plotted in Fig. 2.14b and the evolution of the total phase, i.e., $\Delta\varphi_{XUV} + \Delta\varphi_{at}$ with photon energy is shown in blue.

Often, the contribution of $\Delta\varphi_{at}$ to the measured delay can be calculated independently with sufficient accuracy, such that the phase evolution of the sidebands as a function of XUV photon energy yields the delay in photoemission purely due to the mutual phase of different XUV spectral components [49, 59]. Thus, this measurement delivers information on the phase dispersion of the XUV pulse (train), the so-called *atto-chirp*. Together with knowledge of the XUV power spectrum, this allows for the characterization of the temporal evolution of the APT. This first application of attosecond-PES with APT conferred this method the name RABBITT (*Reconstruction of Attosecond Harmonic Beating by Interference of Two-Photon Transitions* [49]).

Assuming that $\Delta\varphi_{at} \ll \Delta\varphi_{XUV}$ holds for the measurement presented in Fig. 2.14a, b, the observed phase evolution can be attributed to $\Delta\varphi_{XUV}$ predominantly.[5] Thus, as $\Delta\varphi_{XUV}$ indicates the phase difference between the two harmonics contributing to a sideband, and this difference increases linearly (in good approximation, see blue line in Fig. 2.14b) a second-order dispersion (group delay dispersion) can be inferred for the phase of the APT at the tungsten target.

The RABBITT technique can, however, be employed to measure relative delays in the photoemission from different states, via determining $\Delta\varphi_{at}$ under the assumption of a known/constant $\Delta\varphi_{XUV}$. To measure the latter, a material system with known $\Delta\varphi_{at}$ can be used for a reference measurement [52, 60]. Another particularly convenient way to gain information on $\Delta\varphi_{at}$ from a single measurement is to release photoelectrons from different states/bands within the same measurement and, thus, with the same illumination [50, 51, 55]. A measurement according to this concept, performed with the MEGAS beamline is shown in Fig. 2.14c, with the corresponding lineouts through the sidebands plotted in Fig. 2.14d. Here, neon was employed as the nonlinear medium for intracavity HHG, producing photon energies high enough to access electrons both from the valence band and from the more tightly bound 4f core state. The atomic-like behavior of the 4f state ensures photoionization delays among electrons released from this state by different harmonics (for all XUV photon energies employed) small [61] in comparison to the photoionization delays of photoelectrons released from the valence band. In addition, $\Delta\varphi_{XUV}$ is identical for photoelectrons originating from the 4f state and the valence band. While a quantitative interpretation of these data is ongoing and exceeds the scope of this section, with these two pertinent assumptions, two features clearly visible in the sidebands plotted in Fig. 2.14d can be qualitatively explained:

* Firstly, an average delay of approximately 180 as between the valence band and the 4f electrons is observed. This confirms the results of earlier attosecond-PES measurements on tungsten which was mainly attributed to the longer transport time of slower photoelectrons released from the core state [62].
* Secondly, sideband 56 reveals a delay of approximately 160 as with respect to the other sidebands. While the origin of this photoemission delay is still subject to investigations, plausible causes include a resonant excitation to a final state in the band structure of tungsten, in analogy to recent experiments performed on nickel [51].

These experiments demonstrate—for the first time—the applicability of cavity-enhanced HHG for attosecond metrology and the short acquisition times elucidate the benefits of this technology. In our case, in order to access the 4f core state for temporal self-referencing came at the price of a reduction in the photon flux and, therefore, in an increase of the acquisition time from ~1 min to ~3 h, when

---

[5] Here, we make this assumption for illustrative purposes.

using neon instead of argon. However, generating higher photon energies in gases with larger ionization rates might be achievable in the near future by ponderomotive scaling employing longer wavelengths. For instance, with recently demonstrated 100-W-class Tm-fiber frontends [63] generating femtosecond pulses with a central wavelength around 1.95 μm, the wavelength-squared dependence of the HHG cutoff energy would allow the generation of photon energies in excess of 100 eV with peak intensities below 10^{14} W/cm^2. Femtosecond enhancement cavities might constitute a promising route towards counteracting the loss in conversion efficiency due to the longer driving wavelength [64].

References

1. I. Pupeza, T. Eidam, J. Rauschenberger, B. Bernhardt, A. Ozawa, E. Fill, A. Apolonski, T. Udem, J. Limpert, Z.A. Alahmed, A.M. Azzeer, A. Tünnermann, T.W. Hänsch, F. Krausz, Power scaling of a high-repetition-rate enhancement cavity. Opt. Lett. **35**, 2052 (2010)
2. H. Carstens, S. Holzberger, J. Kaster, J. Weitenberg, V. Pervak, A. Apolonski, E. Fill, F. Krausz, I. Pupeza, Large-mode enhancement cavities. Opt. Express **21**, 11606 (2013)
3. I. Pupeza, T. Eidam, J. Kaster, B. Bernhardt, J. Rauschenberger, A. Ozawa, E. Fill, T. Udem, M.F. Kling, J. Limpert, Z.A. Alahmed, A.M. Azzeer, A. Tünnermann, T.W. Hänsch, F. Krausz, Power scaling of femtosecond enhancement cavities and high-power applications, ed. by J.W. Dawson (2011), p. 79141I
4. H. Carstens, N. Lilienfein, S. Holzberger, C. Jocher, T. Eidam, J. Limpert, A. Tünnermann, J. Weitenberg, D.C. Yost, A. Alghamdi, Z. Alahmed, A. Azzeer, A. Apolonski, E. Fill, F. Krausz, I. Pupeza, Megawatt-scale average-power ultrashort pulses in an enhancement cavity. Opt. Lett. **39**, 2595 (2014)
5. C. Jocher, T. Eidam, S. Hädrich, J. Limpert, A. Tünnermann, Sub 25 fs pulses from solid-core nonlinear compression stage at 250 W of average power. Opt. Lett. **37**, 4407 (2012)
6. H. Carstens, M. Högner, T. Saule, S. Holzberger, N. Lilienfein, A. Guggenmos, C. Jocher, T. Eidam, D. Esser, V. Tosa, V. Pervak, J. Limpert, A. Tünnermann, U. Kleineberg, F. Krausz, I. Pupeza, High-harmonic generation at 250 MHz with photon energies exceeding 100 eV. Optica **3**, 366 (2016)
7. N. Lilienfein, H. Carstens, S. Holzberger, C. Jocher, T. Eidam, J. Limpert, A. Tünnermann, A. Apolonski, F. Krausz, I. Pupeza, Balancing of thermal lenses in enhancement cavities with transmissive elements. Opt. Lett. **40**, 843 (2015)
8. S. Holzberger, N. Lilienfein, H. Carstens, T. Saule, M. Högner, F. Lücking, M. Trubetskov, V. Pervak, T. Eidam, J. Limpert, A. Tünnermann, E. Fill, F. Krausz, I. Pupeza, Femtosecond enhancement cavities in the nonlinear regime. Phys. Rev. Lett. **115** (2015)
9. D.R. Carlson, J. Lee, J. Mongelli, E.M. Wright, R.J. Jones, Intracavity ionization and pulse formation in femtosecond enhancement cavities. Opt. Lett. **36**, 2991 (2011)
10. J. Lee, D.R. Carlson, R.J. Jones, Optimizing intracavity high harmonic generation for XUV fs frequency combs. Opt. Express **19**, 23315 (2011)
11. T.K. Allison, A. Cingöz, D.C. Yost, J. Ye, Extreme nonlinear optics in a femtosecond enhancement cavity. Phys. Rev. Lett. **107** (2011)
12. D.C. Yost, A. Cingöz, T.K. Allison, A. Ruehl, M.E. Fermann, I. Hartl, J. Ye, Power optimization of XUV frequency combs for spectroscopy applications [Invited]. Opt. Express **19**, 23483 (2011)
13. I. Pupeza, X. Gu, E. Fill, T. Eidam, J. Limpert, A. Tünnermann, F. Krausz, T. Udem, Highly sensitive dispersion measurement of a high-power passive optical resonator using spatial-spectral interferometry. Opt. Express **18**, 26184 (2010)

14. N. Lilienfein, C. Hofer, M. Högner, T. Saule, M. Trubetskov, V. Pervak, E. Fill, C. Riek, A. Leitenstorfer, J. Limpert, F. Krausz, I. Pupeza, Temporal solitons in free-space femtosecond enhancement cavities. Nat. Photonics **13**, 214–218 (2019)
15. C. Gohle, T. Udem, M. Herrmann, J. Rauschenberger, R. Holzwarth, H.A. Schuessler, F. Krausz, T.W. Hänsch, A frequency comb in the extreme ultraviolet. Nature **436**, 234–237 (2005)
16. R.J. Jones, K.D. Moll, M.J. Thorpe, J. Ye, Phase-coherent frequency combs in the vacuum ultraviolet via high-harmonic generation inside a femtosecond enhancement cavity. Phys. Rev. Lett. **94** (2005)
17. D.C. Yost, T.R. Schibli, J. Ye, Efficient output coupling of intracavity high-harmonic generation. Opt. Lett. **33**, 1099 (2008)
18. A. Cingöz, D.C. Yost, T.K. Allison, A. Ruehl, M.E. Fermann, I. Hartl, J. Ye, Direct frequency comb spectroscopy in the extreme ultraviolet. Nature **482**, 68–71 (2012)
19. A. Ozawa, J. Rauschenberger, C. Gohle, M. Herrmann, D.R. Walker, V. Pervak, A. Fernandez, R. Graf, A. Apolonski, R. Holzwarth, F. Krausz, T.W. Hänsch, T. Udem, High harmonic frequency combs for high resolution spectroscopy. Phys. Rev. Lett. **100** (2008)
20. A. Ozawa, Z. Zhao, M. Kuwata-Gonokami, Y. Kobayashi, High average power coherent vuv generation at 10 MHz repetition frequency by intracavity high harmonic generation. Opt. Express **23**, 15107 (2015)
21. D.C. Yost, T.R. Schibli, J. Ye, Microwatt-level XUV frequency comb via intracavity high harmonic generation, in *Advanced Solid-State Photonics* (OSA, 2009), p. WA1
22. A.K. Mills, T.J. Hammond, M.H.C. Lam, D.J. Jones, XUV frequency combs via femtosecond enhancement cavities. J. Phys. B At. Mol. Opt. Phys. **45**, 142001 (2012)
23. G. Porat, C.M. Heyl, S.B. Schoun, C. Benko, N. Dörre, K.L. Corwin, J. Ye, Phase-matched extreme-ultraviolet frequency-comb generation. Nat. Photonics **12**, 387–391 (2018)
24. C. Corder, P. Zhao, J. Bakalis, X. Li, M.D. Kershis, A.R. Muraca, M.G. White, T.K. Allison, Ultrafast extreme ultraviolet photoemission without space charge. Struct. Dyn. **5**, 054301 (2018)
25. I. Pupeza, S. Holzberger, T. Eidam, H. Carstens, D. Esser, J. Weitenberg, P. Rußbüldt, J. Rauschenberger, J. Limpert, T. Udem, A. Tünnermann, T.W. Hänsch, A. Apolonski, F. Krausz, E. Fill, Compact high-repetition-rate source of coherent 100 eV radiation. Nat. Photonics **7**, 608–612 (2013)
26. K.D. Moll, R.J. Jones, J. Ye, Output coupling methods for cavity-based high-harmonic generation. Opt. Express **14**, 8189 (2006)
27. D. Esser, J. Weitenberg, W. Bröring, I. Pupeza, S. Holzberger, H.-D. Hoffmann, Laser-manufactured mirrors for geometrical output coupling of intracavity-generated high harmonics. Opt. Express **21**, 26797 (2013)
28. L. von der Wense, B. Seiferle, M. Laatiaoui, J.B. Neumayr, H.-J. Maier, H.-F. Wirth, C. Mokry, J. Runke, K. Eberhardt, C.E. Düllmann, N.G. Trautmann, P.G. Thirolf, Direct detection of the ^{229}Th nuclear clock transition. Nature **533**, 47–51 (2016)
29. I. Pupeza, M. Högner, J. Weitenberg, S. Holzberger, D. Esser, T. Eidam, J. Limpert, A. Tünnermann, E. Fill, V.S. Yakovlev, Cavity-enhanced high-harmonic generation with spatially tailored driving fields. Phys. Rev. Lett. **112**, 103902 (2014)
30. J. Weitenberg, P. Rußbüldt, T. Eidam, I. Pupeza, Transverse mode tailoring in a quasi-imaging high-finesse femtosecond enhancement cavity. Opt. Express **19**, 9551 (2011)
31. J. Weitenberg, P. Rußbüldt, I. Pupeza, T. Udem, H.-D. Hoffmann, R. Poprawe, Geometrical on-axis access to high-finesse resonators by quasi-imaging: a theoretical description. J. Opt. **17**, 025609 (2015)
32. A. Ozawa, A. Vernaleken, W. Schneider, I. Gotlibovych, T. Udem, T.W. Hänsch, Non-collinear high harmonic generation: a promising outcoupling method for cavity-assisted XUV generation. Opt. Express **16**, 6233 (2008)
33. C.M. Heyl, S.N. Bengtsson, S. Carlström, J. Mauritsson, C.L. Arnold, A. L'Huillier, Noncollinear optical gating. New J. Phys. **16**, 052001 (2014)
34. M. Louisy, C.L. Arnold, M. Miranda, E.W. Larsen, S.N. Bengtsson, D. Kroon, M. Kotur, D. Guénot, L. Rading, P. Rudawski, F. Brizuela, F. Campi, B. Kim, A. Jarnac, A. Houard, J. Mauritsson, P. Johnsson, A. L'Huillier, C.M. Heyl, Gating attosecond pulses in a noncollinear geometry. Optica **2**, 563 (2015)

35. W.P. Putnam, D.N. Schimpf, G. Abram, F.X. Kärtner, Bessel-Gauss beam enhancement cavities for high-intensity applications. Opt. Express **20**, 24429 (2012)
36. J.P. Marangos, Development of high harmonic generation spectroscopy of organic molecules and biomolecules. J. Phys. B At. Mol. Opt. Phys. **49**, 132001 (2016)
37. I. Jovanovic, M. Shverdin, D. Gibson, C. Brown, High-power laser pulse recirculation for inverse Compton scattering-produced-rays. Nucl. Instrum. Methods Phys. Res. Sect. A **578**, 160–171 (2007)
38. M. Högner, V. Tosa, I. Pupeza, Generation of isolated attosecond pulses with enhancement cavities—a theoretical study. New J. Phys. **19**, 033040 (2017)
39. M. Högner, T. Saule, S. Heinrich, N. Lilienfein, D. Esser, M. Trubetskov, V. Pervak, I. Pupeza, Cavity-enhanced noncollinear high-harmonic generation. Opt. Express **27**, 19675 (2019)
40. M. Högner, T. Saule, I. Pupeza, Efficiency of cavity-enhanced high harmonic generation with geometric output coupling. J. Phys. B At. Mol. Opt. Phys. **52**, 075401 (2019)
41. T. Saule, S. Holzberger, O. de Vries, M. Plötner, J. Limpert, A. Tünnermann, I. Pupeza, Phase-stable, multi-μJ femtosecond pulses from a repetition-rate tunable Ti:Sa-oscillator-seeded Yb-fiber amplifier. Appl. Phys. B **123** (2017)
42. F. Lücking, A. Assion, A. Apolonski, F. Krausz, G. Steinmeyer, Long-term carrier-envelope-phase-stable few-cycle pulses by use of the feed-forward method. Opt. Lett. **37**, 2076 (2012)
43. O. de Vries, T. Saule, M. Plötner, F. Lücking, T. Eidam, A. Hoffmann, A. Klenke, S. Hädrich, J. Limpert, S. Holzberger, T. Schreiber, R. Eberhardt, I. Pupeza, A. Tünnermann, Acousto-optic pulse picking scheme with carrier-frequency-to-pulse-repetition-rate synchronization. Opt. Express **23**, 19586 (2015)
44. J. Weitenberg, T. Saule, J. Schulte, P. Rusbuldt, Nonlinear pulse compression to sub-40 fs at μJ pulse energy by multi-pass-cell spectral broadening. IEEE J. Quantum Electron. **53**, 1–4 (2017)
45. T. Saule, S. Heinrich, J. Schötz, N. Lilienfein, M. Högner, O. de Vries, M. Plötner, J. Weitenberg, D. Esser, J. Schulte, P. Russbueldt, J. Limpert, M.F. Kling, U. Kleineberg, I. Pupeza, High-flux ultrafast extreme-ultraviolet photoemission spectroscopy at 18.4 MHz pulse repetition rate. Nat. Commun. **10**, 458 (2019)
46. T. Saule, M. Högner, N. Lilienfein, O. de Vries, M. Plötner, V.S. Yakovlev, N. Karpowicz, J. Limpert, I. Pupeza, Cumulative plasma effects in cavity-enhanced high-order harmonic generation in gases. APL Photonics **3**, 101301 (2018)
47. M. Lewenstein, P. Balcou, M.Y. Ivanov, A. L'Huillier, P.B. Corkum, Theory of high-harmonic generation by low-frequency laser fields. Phys. Rev. A **49**, 2117–2132 (1994)
48. P.M. Paul, Observation of a train of attosecond pulses from high harmonic generation. Science **292**, 1689–1692 (2001)
49. H.G. Muller, Reconstruction of attosecond harmonic beating by interference of two-photon transitions. Appl. Phys. B **74**, s17–s21 (2002)
50. C. Chen, Z. Tao, A. Carr, P. Matyba, T. Szilvási, S. Emmerich, M. Piecuch, M. Keller, D. Zusin, S. Eich, M. Rollinger, W. You, S. Mathias, U. Thumm, M. Mavrikakis, M. Aeschlimann, P.M. Oppeneer, H. Kapteyn, M. Murnane, Distinguishing attosecond electron–electron scattering and screening in transition metals. Proc. Natl. Acad. Sci. **114**, E5300–E5307 (2017)
51. Z. Tao, C. Chen, T. Szilvasi, M. Keller, M. Mavrikakis, H. Kapteyn, M. Murnane, Direct time-domain observation of attosecond final-state lifetimes in photoemission from solids. Science **353**, 62–67 (2016)
52. R. Locher, L. Castiglioni, M. Lucchini, M. Greif, L. Gallmann, J. Osterwalder, M. Hengsberger, U. Keller, Energy-dependent photoemission delays from noble metal surfaces by attosecond interferometry. Optica **2**, 405 (2015)
53. M. Lucchini, L. Castiglioni, L. Kasmi, P. Kliuiev, A. Ludwig, M. Greif, J. Osterwalder, M. Hengsberger, L. Gallmann, U. Keller, Light-matter interaction at surfaces in the spatiotemporal limit of macroscopic models. Phys. Rev. Lett. **115** (2015)
54. L. Cattaneo, J. Vos, M. Lucchini, L. Gallmann, C. Cirelli, U. Keller, Comparison of attosecond streaking and RABBITT. Opt. Express **24**, 29060 (2016)

55. M. Isinger, R.J. Squibb, D. Busto, S. Zhong, A. Harth, D. Kroon, S. Nandi, C.L. Arnold, M. Miranda, J.M. Dahlström, E. Lindroth, R. Feifel, M. Gisselbrecht, A. L'Huillier, Photoionization in the time and frequency domain. Science **358**, 893–896 (2017)
56. S. Heinrich, T. Saule, M. Högner, Y. Cui, V.S. Yakovlev, I. Pupeza, U. Kleineberg, Attosecond intra-valence band dynamics and resonant-photoemission delays in W(110). Nat. Commun. **12**, 3404 (2021)
57. S. Hüfner, *Photoelectron Spectroscopy: Principles and Applications*, 3rd edn. Advanced Texts in Physics (Springer, 2003)
58. L. Miaja-Avila, C. Lei, M. Aeschlimann, J.L. Gland, M.M. Murnane, H.C. Kapteyn, G. Saathoff, Laser-assisted photoelectric effect from surfaces. Phys. Rev. Lett. **97** (2006)
59. Y. Mairesse, Attosecond synchronization of high-harmonic soft X-rays. Science **302**, 1540–1543 (2003)
60. M. Ossiander, F. Siegrist, V. Shirvanyan, R. Pazourek, A. Sommer, T. Latka, A. Guggenmos, S. Nagele, J. Feist, J. Burgdörfer, R. Kienberger, M. Schultze, Attosecond correlation dynamics. Nat. Phys. **13**, 280–285 (2016)
61. M. Ossiander, J. Riemensberger, S. Neppl, M. Mittermair, M. Schäffer, A. Duensing, M.S. Wagner, R. Heider, M. Wurzer, M. Gerl, M. Schnitzenbaumer, J.V. Barth, F. Libisch, C. Lemell, J. Burgdörfer, P. Feulner, R. Kienberger, Absolute timing of the photoelectric effect. Nature **561**, 374–377 (2018)
62. A.L. Cavalieri, N. Müller, T. Uphues, V.S. Yakovlev, A. Baltuška, B. Horvath, B. Schmidt, L. Blümel, R. Holzwarth, S. Hendel, M. Drescher, U. Kleineberg, P.M. Echenique, R. Kienberger, F. Krausz, U. Heinzmann, Attosecond spectroscopy in condensed matter. Nature **449**, 1029–1032 (2007)
63. C. Gaida, T. Heuermann, M. Gebhardt, E. Shestaev, T.P. Butler, D. Gerz, N. Lilienfein, P. Sulzer, M. Fischer, R. Holzwarth, A. Leitenstorfer, I. Pupeza, J. Limpert, High-power frequency comb at 2 μm wavelength emitted by a Tm-doped fiber laser system. Opt. Lett. **43**, 5178 (2018)
64. A.D. Shiner, C. Trallero-Herrero, N. Kajumba, H.-C. Bandulet, D. Comtois, F. Légaré, M. Giguère, J.-C. Kieffer, P.B. Corkum, D.M. Villeneuve, Wavelength scaling of high harmonic generation efficiency. Phys. Rev. Lett. **103** (2009)

Chapter 3
Next-Generation Enhancement Cavities for Attosecond Metrology—An Outlook

Precisely tracing complex attosecond dynamics might benefit from—or even necessitate—isolated attosecond pulses (IAP) rather than attosecond pulse trains [1–3]. The conceptually most straightforward approach for the generation of IAP is the use of few-cycle driving pulses so short that the harmonic emission is restricted to a single half-cycle of the driving field. However, the bandwidth necessary to support such short pulses poses severe requirements to the optics in such experiments in general, and to enhancement cavity mirrors in particular. In fact, so far, for the wavelength range around 1 μm EC mirrors supporting single-cycle pulses and, simultaneously, a power enhancement of at least 10 have not been demonstrated.

Alternatively, gating mechanisms have been explored that confine the harmonic emission (along a given direction) to a single half-cycle of the driving field. The latter option relaxes the requirements to the driving source and the bandwidth of the optics. However, the efficiency of gating mechanisms drastically decreases with increasing pulse duration, such that in this case, too, driving pulses as short as possible are highly desirable. A second prerequisite for the pulse-to-pulse reproducible generation of IAP is the waveform stability of the driving light field.

Section 3.1 summarizes our results towards the two goals mentioned above—the enhancement of waveform-stable pulses and scaling the optical bandwidth of ECs. Based on these results, we have conducted studies for the identification of a gating method for the generation of IAP with ECs. Section 3.2 discusses this study.

Our studies of intracavity nonlinearities have spawned an intriguing result, namely the first demonstration of temporal dissipative solitons in free-space ECs. The nonlinear dynamics of femtosecond pulses in this regime might offer novel potential for HHG (and for the generation of IAP). These results are summarized in Sect. 3.3.

I. Pupeza, *Passive Optical Resonators for Next-Generation Attosecond Metrology*, SpringerBriefs in Physics, https://doi.org/10.1007/978-3-030-92972-5_3

3.1   Passive Enhancement of Few-Cycle, Waveform-Stable Pulses

3.1.1   Enhancement Cavities for Zero-Offset-Frequency Pulse Trains [4]

During propagation in the resonator, the carrier-envelope phase (CEP) of the circulating pulse is modified by mechanisms such as dispersion along the beam path, Gouy phase shifts, and reflections off multilayer mirrors. To achieve waveform-stable circulating pulses, the net CEP shift per roundtrip must be zero. In this paper, we demonstrated that the design of broadband multilayer mirrors can be adapted such that the roundtrip CEP slippage can be set to zero with sufficient accuracy for the enhancement of 30-fs pulses by a factor of 200, see Fig. 3.1.

3.1.2   Enhancement Cavities for Few-Cycle Pulses [5]

The challenge of building a very broadband "standard-approach" EC, i.e., exhibiting a flat phase of the response function, lies in the fact that any deviation from a flat phase is amplified in the steady state by a factor close to the power enhancement (assuming good transverse modematching). While our studies[1] indicated that in theory dielectric multilayer mirror designs with a sufficiently flat phase are feasible even for 10-fs pulses, we found that the realization of these complex designs was beyond feasibility for commercial optics.

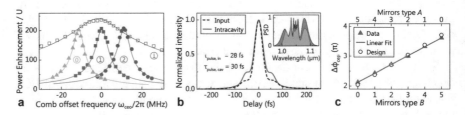

Fig. 3.1 For this experiment, mirrors of two types (A and B) were produced, exhibiting very similar reflectivity and dispersion, but different CEP shifts. **a** Power enhancement corrected for the spatial overlap (U) versus offset frequency of the seeding comb for a cavity employing 0, 1, or 2 mirrors of type B (indicated by the numbers). Solid symbols: broadband input pulses (30 fs). Open symbols: input pulses of reduced bandwidth, illustrating the dependence of the optimum input comb offset frequency on the spectrum. Solid lines: numerical model. **b** Intensity autocorrelation trace of the pulses in case of optimum comb parameters. Inset: corresponding normalized power spectral density (PSD) with 100-nm intracavity bandwidth at −10 dB. **c** Retrieved CEP shift upon reflection off five cavity mirrors and calculated values from the coating design. Adapted with permission from [4] © The Optical Society

---

[1] These studies were mainly carried out by M. Trubetskov.

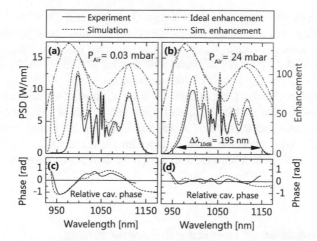

Fig. 3.2 Intracavity spectra, and spectral enhancement **a** in vacuum and **b** with 24 mbar of air pressure. The spectral enhancement is shown for "ideal" mirrors with a flat phase (but with a limited spectral coverage of high reflectivity, corresponding to the one from the experiment) as well as for the measured mirror phase. **c, d** Measured and calculated phase of the cavity response, for vacuum and for 24 mbar of air. Adapted with permission from [5] © The Optical Society

To investigate the shortest pulse duration achievable at high-power enhancement, we characterized the spectral phase of several mirrors designed according to the criterion of flat phase over the broadest possible bandwidth, with a multi-pass spatial-spectral interferometry setup. By combining mirrors from several coating runs to achieve a phase as flat as possible, we demonstrated the enhancement of pulses comprising only 5.4 cycles, spanning a spectrum of almost 200 nm (at −10 dB) around a central wavelength of 1050 nm, to kW-level average powers (Fig. 3.2a). By adding 24 mbar of air to the beam path, the roundtrip phase was adjusted even closer to zero, as shown in Fig. 3.2b.

These results had a twofold implication for further work on EC-HHG. Firstly, they provided a realistic minimum pulse duration achievable with state-of-the-art mirror coating technology, as an important basis for further studies towards the generation of IAP. Secondly, as both the HHG efficiency and the achievable XUV photon energies scale with decreasing pulse duration [6, 7], they demonstrated the potential for further scaling these parameters.

3.2 Toward Intracavity Gating for the Generation of Isolated Attosecond Pulses

Having demonstrated the necessary prerequisites of short, waveform-stable circulating pulses, we set out to identify suitable methods for confining the harmonic emission (in a given spatial direction) to a single attosecond XUV burst, i.e., cavity-enhanced gating methods for the generation of isolated attosecond pulses (IAP). The following sections summarize this work. The first step was a comprehensive theoretical study of several gating methods that seemed promising in the context of a femtosecond EC. Having identified *transverse mode gating* (TMG) as the method of choice (explained below), we carried out a set of proof-of-principle experiments, demonstrating all necessary ingredients for the generation of IAP. Finally, an interferometric position tracking method is presented, allowing few-attosecond-precision delay determination between a NIR field and an attosecond XUV burst, as a solution for time-resolved measurements in which the two pulses travel along separate pathways.

3.2.1   Generation of Isolated Attosecond Pulses with Enhancement Cavities—A Theoretical Study [8]

In a comprehensive theoretical study, we investigated the suitability of gating methods for the generation of IAP using ECs. For this study we assumed pulse parameters experimentally achieved with Yb-based, high-repetition-rate femtosecond laser technology (0.7 μJ, 5 cycles, see previous chapter). It is worth noting that modern pulse compression schemes afford significantly higher pulse energies for such pulse durations [9], considerably relaxing the conditions for EC-based IAP generation.

In our study, we considered amplitude gating, ionization gating, polarization gating, multi-color collinear superposition (such as double optical gating), angular streaking (attosecond lighthouse, non-collinear optical gating). We identified polarization gating and TMG, an implementation of angular streaking, as the most promising methods, with the latter promising a higher overall conversion efficiency. Our study predicted the feasibility of IAPs with photon energies around 100 eV and a photon flux of at least 10^8 photons/s at repetition rates of 10 MHz and higher, starting with the above-mentioned pulse parameters. This triggered first experiments on the feasibility of a TMG-ready EC geometry, described in the next paragraph.

3.2.2   Tailoring the Transverse Mode of a High-Finesse Optical Resonator with Stepped Mirrors [10]

Transverse mode gating relies on tailoring the transverse mode resonant in the EC by means of monolithic optics in order to achieve a wavefront rotation (WFR) in the region of a tight focus. Specifically, we have introduced a temporal offset between the two lobes of the TEM_{01} mode of a standard bowtie ring resonator by means of a mirror exhibiting a step profile, see Fig. 3.3a. In the case of pulsed radiation, the wavefront modification results in a WFR upon focusing, and the temporal offset of

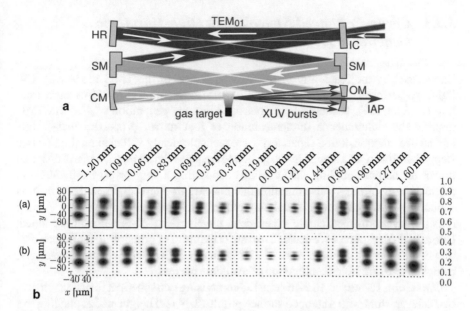

Fig. 3.3 a Principle of TMG: the TEM_{01} transverse mode of a bowtie ring resonator is modified by means of a stepped mirror (SM) to introduce a delay between the two lobes of the mode. The modification of the mode is reversed after the focus with a second SM. In case of pulsed excitation of this mode, a WFR at the focus leads to the spatially dispersed emission of the XUV bursts generated with each half-cycle of the driving field. If the separation is strong enough, one of these beamlets can be efficiently separated at the output coupling mirror (OM). IC: input coupler, HR: highly reflective mirror. **b** Measured (upper panel) and simulated (lower panel) transverse intensity profiles (x–y-plane) of the TMG mode at the cavity focus, for different positions along the optical axis, for CW excitation. Adapted with permission from [10] © IOP Publishing

the two lobes can be reversed with negligible diffraction losses if a second, identical stepped mirror is arranged in a nearly imaging configuration to the first one, see Fig. 3.3a.

To demonstrate that such a mode can be excited with low diffraction losses, we set up a proof-of-principle EC, and used a single-frequency continuous-wave (CW) laser (at 1064 nm) to seed it. Figure 3.3b shows the intensity profile measured at the focus, agreeing very well with the numerically simulated one and, in particular, exhibiting the expected on-axis maximum at the focus. We obtained a finesse of 652, corresponding to a round-trip loss of 0.36%. The power enhancement was 97, at a spatial overlap with the input beam of 37%. Theoretically, the spatial overlap can be improved to 83% by means of cylindrical lenses and phase masks placed before the EC, which would increase the enhancement to 217. Importantly, the EC was realized in a geometry which would have allowed both for peak intensities high enough to generate IAP with photon energies in the 100-eV range, and spot sizes on the mirrors large enough to avoid damage with state-of-the-art cavity mirrors. This proof of principle triggered the experiments with pulsed excitation reported in the next section.

3.2.3   Cavity-Enhanced Noncollinear High-Harmonic Generation [11]

Encouraged by the feasibility of low-loss excitation of the TMG mode with CW light, we performed proof-of-principle experiments with the MEGAS infrastructure (cf. Sect. 2.4). It should be emphasized that the generation of IAP via TMG requires the simultaneous implementation of four mirror properties that—while having been demonstrated separately—are technologically still challenging: (i) two stepped mirrors with a step height adjusted to the pulse duration, (ii) a pierced/slotted mirror for output coupling, (iii) coatings supporting a very broad bandwidth—in particular supporting a pulse duration that matches the step height, and (iv) mirrors supporting waveform-stable circulating pulses. Unfortunately, three mirror coating runs for our proof-of-principle experiment, targeting sub-20-fs pulses, have failed. Due to limited resources we decided to restrict the proof of principle to the demonstration of the most important feature of TMG, namely HHG with a rotating wavefront, using the most suitable combination of available mirrors.

We set up a 10-mirror, 16.3-m-long EC comprising two spherical focusing mirrors, two delay mirrors with a stepped surface profile (2.34 μm height, corresponding to a delay of 2.57 cycles at a wavelength of 1025 nm), an input coupler with a reflectivity of 97.2% and five highly reflective plane mirrors. The EC was seeded by the MEGAS frontend, in this experiment providing 2.7-μJ, 38-fs pulses with a repetition rate of 18.4 MHz and spectrally centered at 1030 nm. The highly reflective mirror coatings supported ~30-fs pulses. To afford high powers, we operated the EC close to the inner stability edge. By selecting a proper combination of mirror coatings, we obtained a cavity with a preferred comb-offset frequency of zero, i.e. supporting waveform-stable circulating pulses. The size of the focal spot was $w_{0x} \times w_{0y} = 17.2 \times 11.8\,\mu m^2$. We measured 80-μJ intracavity pulses with a spectrum centered at 1025 nm in the empty cavity, at a finesse of 188 (round-trip loss of 0.5%).

Imaging the focal region through a variable band-pass filter revealed the expected WFR, see Fig. 3.4a. Furthermore, we generated and detected harmonics in argon and neon, with photon energies up to 60 and 120 eV, respectively. Both the WFR and the harmonic cutoffs were in good agreement with the values predicted by theory, validating the basic concept. Extrapolating these results to an intracavity pulse duration of 17.5 fs—which is, in principle, feasible [5]—predicts the generation of IAP with a contrast ratio approaching 100, see Fig. 3.4b.

In conclusion, we have demonstrated all necessary ingredients for the efficient generation of isolated XUV attosecond pulses at repetition rates of several tens of MHz employing transverse mode gating in femtosecond enhancement cavities. The failure to directly demonstrate the emergence of actual IAP was due to the technically challenging combination of the requirements (i)–(iv) to the EC mirrors, as discussed above. However, based on the fact that mirrors exhibiting each of the properties (i)–(iv) have been previously demonstrated, we foresee that with dedicated resources this novel scheme for the generation of IAP can be readily implemented.

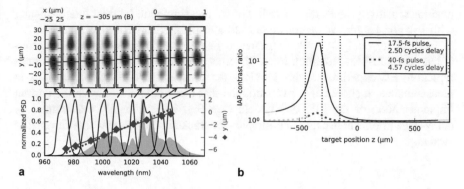

Fig. 3.4 **a** Wavefront rotation observed close to the focus of the TMG cavity. Top: image of the transverse intensity profiles of the cavity mode, at a fixed plane 305 μm in front of the focus, bandpass-filtered for different wavelengths. The blue lines mark the position (solid) and $1/e^2$-intensity width (dotted) of Gaussian functions fitted to the central lobe of the horizontally integrated profile. Bottom: intracavity spectrum (gray) and, for each depicted profile, the corresponding spectrum transmitted through the bandpass filter (black, normalized). The diamonds mark the central lobe positions and corresponding central wavelengths of the filtered spectra. A linear fit (blue solid line) of the lobe position versus frequency yields a spatial dispersion of 44 μm/PHz, compared to a theoretical value of 53 μm/PHz (blue dotted line). **b** Calculated on-axis inter-burst contrast ratio versus gas target position using 17.5-fs pulses and an accordingly chosen delay introduced by the stepped mirrors (solid black line), and with the 40-fs pulses form our experiment (dashed red line), in both cases computed for the 79th harmonic in neon at a peak intensity of 3.0×10^{14} W/cm^2. The grey continuations of the curves identify when pulse energies >80 μJ are needed to reach the corresponding peak intensities. The vertical dotted lines mark gas target positions from the experiment (see manuscript for details). Adapted with permission from [11] © The Optical Society

3.2.4   Interferometric Delay Tracking for Low-Noise Mach–Zehnder-Type Scanning Measurements [12]

In contrast to the generation of attosecond pulse trains with the fundamental TEM$_{00}$ mode (see Sect. 2.4), in the TMG scheme, the NIR light leaving the EC through the opening in the output coupling mirror is not necessarily suited as a pump (or probe) in attosecond PES experiments. Its power is low due to the on-axis intensity minimum on the output coupling mirror and the duration of the output coupled NIR pulses is at best that of the circulating light. Therefore, the coherent superposition of the IAP with waveform-stable few-cycle pulses generated somewhere else in the system (e.g., directly from the MEGAS oscillator, see Fig. 2.9) might be desirable. To this end, an interferometric readout of the relative delay between the two pulses might be very helpful.

We modified a commercial laser-based positioning system to enable ultrafast (up to 1 MHz) interferometric delay tracking for Mach–Zehnder-type setups. The system was realized with a 1550-nm pilot beam. Modulation and subsequent demodulation of the resulting interference signal at two mutually phase-shifted and different frequencies allows the construction of a quadrature signal. This permits both a constantly high position sensitivity, and a directionality of the measurement. In a test interferometer with 1-m arm length, we demonstrated sub-attosecond delay tracking precision.

Importantly, our system allows for mechanically chopping one arm of the interferometer at multi-kHz repetition rates. Together with the multi-MHz repetition rates typical to EC experiments, this provides the prospect of high-sensitivity lock-in measurements. It should also be mentioned that the timing accuracy in the current implementation of the MEGAS system (see Fig. 2.9) is mainly constrained by delay stage jitter, which can—in principle—be tracked with this new interferometric system.

3.3   Solitons in Free-Space Femtosecond Enhancement Cavities [13]

One of the most intriguing and maybe important results of our preoccupation with femtosecond ECs in the nonlinear regime has been the first demonstration of temporal dissipative solitons (TDS) in free-space, macroscopic ECs. TDS are wave packets circulating in nonlinear optical resonators in a regime of self-compressed, self-stabilizing propagation, ensured by a balance between linear and nonlinear phase shifts. Owing to these highly desirable properties, TDS have been harnessed for the generation of ultrashort pulses and frequency combs in various active and passive laser architectures. Most notably, in recent years numerous studies have addressed the formation of *cavity solitons* (CS) in passive fibre resonators and microresonators. In our work, we demonstrated the formation of CS in a free-space enhancement cavity with a Kerr nonlinearity and a spectrally tailored finesse, see Fig. 3.5a. Locking a 100-MHz-repetition-rate train of 350-fs pulses spectrally centered at 1030 nm to this CS state, we generated an intracavity 37-fs $sech^2$-shaped pulse with a peak-power enhancement of 3200, see Fig. 3.5b. This novel platform of ultrashort pulses exhibits three unique properties:

- The tailored input coupler reflectivity leads to an unprecedented input coupling efficiency for CS. In particular, the frequency components generated inside of the EC experience exclusively reflections off high-reflectivity mirrors, while the enhancement of the input spectrum can be tailored close to impedance matching.
- In addition to the high-frequency noise filtering properties typical to linear ECs, we observed a low-frequency intensity-noise suppression.
- Free-space ECs afford power scalability properties unrivaled by other laser architectures, owing to their simplicity, scalability of the mode sizes on the optics and separation of wave-guiding properties and nonlinearity. In fact, the soliton peak power of 125 MW demonstrated in our proof-of-principle experiment exceeded the critical power for self-focusing in the Kerr medium material (sapphire) by a factor of 40.

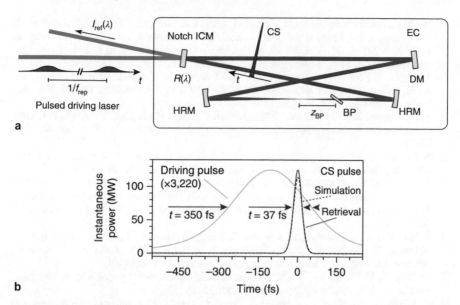

Fig. 3.5 **a** Principle of the generation of CS in free-space ECs. Seeding pulses impinge on an input coupling mirror (ICM) with a non-zero transmission only over their spectrum. The interplay of the Kerr nonlinearity in a sapphire Brewster plate (BP) and the negatively dispersive mirror (DM) leads to the formation of a compressed circulating CS. HRM: highly reflective mirror. **b** Intensity envelopes (instantaneous power) of the seeding pulses and the CS as measured by FROG, together with simulation results for the CS. Adapted with permission from [13] © Springer Nature

These properties promise a vast potential for applications including spatiotemporal filtering and compression of ultrashort pulses, and cavity-enhanced nonlinear frequency conversion. In the particular context of attosecond metrology, this scheme could be developed towards an external pulse compression method, as a noise-filtering alternative to state-of-the-art multi-pass-cell-based temporal compression (see Sect. 2.4). Furthermore, it could provide a way to overcome plasma-nonlinearity-induced intensity clamping in EC-HHG (see Sect. 2.2)—either by adding a Kerr nonlinearity in a bulk or by using the plasma nonlinearity of the gas target itself [7]. The soliton dynamics might afford an increased robustness against phase variations of the EC mirrors, enabling the enhancement of shorter pulses than with standard-approach ECs, improving the HHG efficiency and facilitating IAP generation.

References

1. M.I. Stockman, M.F. Kling, U. Kleineberg, F. Krausz, Attosecond nanoplasmonic-field microscope. Nat. Photonics **1**, 539–544 (2007)

2. B. Förg, J. Schötz, F. Süßmann, M. Förster, M. Krüger, B. Ahn, W.A. Okell, K. Wintersperger, S. Zherebtsov, A. Guggenmos, V. Pervak, A. Kessel, S.A. Trushin, A.M. Azzeer, M.I. Stockman, D. Kim, F. Krausz, P. Hommelhoff, M.F. Kling, Attosecond nanoscale near-field sampling. Nat. Commun. **7**, 11717 (2016)
3. J. Schötz, B. Förg, M. Forster, W.A. Okell, M.I. Stockman, F. Krausz, P. Hommelhoff, M.F. Kling, Reconstruction of nanoscale near fields by attosecond streaking. IEEE J. Sel. Top. Quantum Electron. **23**, 77–87 (2017)
4. S. Holzberger, N. Lilienfein, M. Trubetskov, H. Carstens, F. Lücking, V. Pervak, F. Krausz, I. Pupeza, Enhancement cavities for zero-offset-frequency pulse trains. Opt. Lett. **40**, 2165 (2015)
5. N. Lilienfein, C. Hofer, S. Holzberger, C. Matzer, P. Zimmermann, M. Trubetskov, V. Pervak, I. Pupeza, Enhancement cavities for few-cycle pulses. Opt. Lett. **42**, 271 (2017)
6. I. Pupeza, S. Holzberger, T. Eidam, H. Carstens, D. Esser, J. Weitenberg, P. Rußbüldt, J. Rauschenberger, J. Limpert, T. Udem, A. Tünnermann, T.W. Hänsch, A. Apolonski, F. Krausz, E. Fill, Compact high-repetition-rate source of coherent 100 eV radiation. Nat. Photonics **7**, 608–612 (2013)
7. S. Holzberger, N. Lilienfein, H. Carstens, T. Saule, M. Högner, F. Lücking, M. Trubetskov, V. Pervak, T. Eidam, J. Limpert, A. Tünnermann, E. Fill, F. Krausz, I. Pupeza, Femtosecond enhancement cavities in the nonlinear regime. Phys. Rev. Lett. **115** (2015)
8. M. Högner, V. Tosa, I. Pupeza, Generation of isolated attosecond pulses with enhancement cavities—a theoretical study. New J. Phys. **19**, 033040 (2017)
9. K. Fritsch, M. Poetzlberger, V. Pervak, J. Brons, O. Pronin, All-solid-state multipass spectral broadening to sub-20 fs. Opt. Lett. **43**, 4643 (2018)
10. M. Högner, T. Saule, N. Lilienfein, V. Pervak, I. Pupeza, Tailoring the transverse mode of a high-finesse optical resonator with stepped mirrors. J. Opt. **20**, 024003 (2018)
11. M. Högner, T. Saule, S. Heinrich, N. Lilienfein, D. Esser, M. Trubetskov, V. Pervak, I. Pupeza, Cavity-enhanced noncollinear high-harmonic generation. Opt. Express **17** (2019)
12. W. Schweinberger, L. Vamos, J. Xu, C. Baune, S. Rode, Interferometric delay tracking for low-noise Mach-Zehnder-type scanning measurements. Opt. Express **27**, 4789–4798 (2019)
13. N. Lilienfein, C. Hofer, M. Högner, T. Saule, M. Trubetskov, V. Pervak, E. Fill, C. Riek, A. Leitenstorfer, J. Limpert, F. Krausz, I. Pupeza, Temporal solitons in free-space femtosecond enhancement cavities. Nat. Photonics **13**, 214–218 (2019)

Printed in the United States
by Baker & Taylor Publisher Services